APPLIED HUMAN FACTORS IN AVIATION MAINTENANCE

Dedicated to all the aviation maintenance professionals who do not compromise safety—around the world and around the clock

Applied Human Factors in Aviation Maintenance

MANOJ S. PATANKAR
Saint Louis University, USA

JAMES C. TAYLOR
Aviation Consultant, Los Altos, USA

ASHGATE

Published by
Ashgate Publishing Limited
Gower House
Croft Road
Aldershot
Hampshire GU11 3HR
England

Ashgate Publishing Company
Suite 420
101 Cherry Street
Burlington, VT 05401-4405
USA

Ashgate website: http://www.ashgate.com

British Library Cataloguing in Publication Data
Patankar, Manoj S.
Applied human factors in aviation maintenance
1.Airplanes - Maintenance and repair - Management
2.Airplanes - Maintenance and repair - Safety measures
3.Aeronautics - Human factors
I.Title II.Taylor, James C. (James Chapman), 1937-
629.1'346'0684

Library of Congress Cataloging-in-Publication Data
Patankar, Manoj S., 1968-
Applied human factors in aviation maintenance /
by Manoj S. Patankar and James C. Taylor.
p. cm.
Includes bibliographical references and index.
ISBN 978-0-7546-1940-6
1. Airplanes--Maintenance and repair. 2. Aeronautics--Human factors. 3. Aeronautics--Safety measures. I. Taylor, James C. (James Chapman), 1937- II. Title

TL671.9.P3797 2004
629.134'6--dc22

2004010325

ISBN 978 0 7546 1940 6

Reprinted 2007

Printed and bound in Great Britain by MPG Books Ltd, Bodmin, Cornwall

Contents

List of Examples

List of Figures

List of Tables

Preface

This book serves as a companion to *Risk Management and Error Reduction in Aviation Maintenance* published by us in 2004. As such, we focus on the application of human factors. Emphasis is placed on data from the NASA Aviation Safety Reporting System (ASRS) to illustrate how human factors principles and error reduction techniques should be applied to minimize error-inducing conditions in the future as well as to minimize the impact of past errors.

Considering the global awareness of human performance issues affecting maintenance personnel, there is enough evidence in the ASRS reports to establish that systemic problems such as impractical maintenance procedures, inadequate training, and safety versus profit challenge continue to contribute toward latent failures. Although a handful of error mitigation techniques have been used to minimize such latent failures, the sustained use of such error mitigation techniques has been marred by factors including, but not limited to, the following: low mechanics' trust in their management, inconsistent professionalism among the mechanics, and limited regulatory and corporate resources allocated to safety issues.

In this book, we review the key concepts discussed in our previous book and provide specific guidance to help both mechanics as well as managers implement changes and sustain those changes.

Chapter 1 provides a review of the key concepts such as the Hawkins-Ashby model, different levels of risk, various error-mitigation techniques, and a set of personal habits that would raise the maintainers' professionalism. In Chapter 2, we discuss the transition from awareness to implementation—actually making some changes—at both the individual level as well as the organizational level. The role of labor unions is also discussed. In Chapters 3 and 4, we illustrate the use of two risk management tools: pre-task scorecard and post-task scorecard. At the individual level, these tools can be used to quantify the level of risk involved in performing a particular maintenance task. Similarly, these tools can be used by organizations to measure the risk at a team, shift, or organizational unit level. In Chapter 5, we discuss the two most critical challenges in aviation maintenance: professionalism and trust. The issue of interpersonal trust mainly concerns an individual's belief that his/her supervisor will choose safety over profit. In the collective aftermath of corporate financial scandals, the September 11 tragedy, the Iraq war, and the SARS epidemic, the choice between safety and performance/profit has become even more important. Consequently, the issues of professionalism and interpersonal trust are more significant than ever before. Finally, in Chapter 6, we discuss the challenges in collecting, analyzing, and communicating safety-critical data. Most importantly, the issues of confidentiality and anonymity are discussed.

Ultimately, the application of human factors principles needs to take place across multiple dimensions: in colleges and universities, in maintenance organizations, in regulatory organizations, and among accident investigators. Together, we need to practice, preach, and improve our understanding of human factors in maintenance.

We thank you very much for your interest in our ideas and our research results and we hope that you will find this book useful in the improvement of your practice of human factors in aviation maintenance. As always, we encourage you to reflect upon your personal experiences as you read this book and try to be more proactive in your quest for safety.

Respectfully,

Manoj S. Patankar
Chesterfield, MO

James C. Taylor
Los Altos, CA

Acknowledgments

Research in aviation maintenance is not possible without generous help from numerous practitioners—the hardworking and genuinely caring individuals like you. Over the years, literally thousands of mechanics, inspectors, managers, maintenance engineers, and support personnel have entrusted us with very valuable information. They have taken the time to respond to survey questionnaires, participate in interviews, and allowed us to observe their work practices. We are eternally grateful to them for their hospitality and openness.

We would like to extend our special gratitude to the following individuals for their relentless support and encouragement: David Driscoll, Scott Gilliland, John Goglia, Jay Hiles, Barbara Kanki, Kevin Lynch, Robert and Gordon Mudge, Ken Peartree, Michael Peate, Zoe Sexhus, Sudhir Gopinathan, Robert Thomas III, and Jean Watson.

We would also like to acknowledge the following universities for their support: Saint Louis University—Parks College of Engineering and Aviation, San Jose State University—Department of Aviation and the Institutional Review Board, and Santa Clara University—School of Engineering.

We thank John Hindley and his staff at Ashgate Publishing Limited for providing excellent help in bringing this project to fruition in a timely manner.

Finally, we thank our family members for their support, and encouragement to pursue this project: Kirsten and Sanjeevani Patankar and Ellen Jo Baron.

Chapter 1

Implementing Human Factors in Maintenance: Individual, Organizational and Collegiate Perspectives

Instructional Objectives

Upon completing this chapter, you should be able to accomplish the following:

1. Describe how mechanics were able to apply their human factors training to change their individual behavior.
2. Discuss specific ways to implement human factors principles at the individual level.
3. Describe how certain organizations were able to garner support to implement and sustain human factors/safety initiatives.
4. Explain how human factors training could be incorporated in a collegiate aviation maintenance curriculum.
5. Discuss the factors that contribute toward the success of human factors programs.
6. Discuss the factors that contribute toward the failure of human factors programs.
7. Discuss the influence of regulatory initiatives such as JAR 66, and 145, as well as FAA's AC 120-16D and 120-79 on application of human factors principles in aviation maintenance.

Introduction

In this chapter, we will review some of the topics typically covered in most awareness-level courses and focus on the implementation of those topics, from a behavioral perspective, at three levels: individual, organizational, and collegiate. We will use specific examples from NASA's Aviation

Safety Reporting System (ASRS) reports as well as the Joint Aviation Authorities' JAR 66 and 145 requirements to illustrate our recommendations. Additionally, we will discuss the implications of the Federal Aviation Administration's Advisory Circulars 120-16D *Air Carrier Maintenance Programs* and 120-79 *Developing and Implementing a Continuing Analysis and Surveillance System* from the perspectives of organizational change and commitment.

A Review of Human Factors Awareness Topics

Typical human factors training programs in the aviation maintenance industry have been dominated by awareness-level training, also called 'third generation MRM programs' (Patankar & Taylor, 2004, pp. 66-77). The curriculum for such a training program typically included the following components:

1. *Dirty Dozen elements*: Lack of communication, complacency, lack of knowledge, distraction, lack of teamwork, fatigue, lack of resources, pressure, lack of assertiveness, stress, lack of awareness, and norms. Safety nets associated with each of these elements are also discussed.
2. *Accident case analysis*: One or more exercises designed to illustrate how a chain of events (at times each event is a minor deviance) can lead to disastrous consequences.
3. *Organization-specific problem*: Focus on a particular problem that the organization wants to rectify immediately. Examples of such problems include shift turnovers, logbook errors, ground damage, or lost-time injuries.
4. *Interactive exercises*: Typically, the training also included at least one interactive exercise to illustrate concepts such as the value of teamwork or hazards of verbal turnovers.

In general, we have observed that such training programs have been very effective in raising the awareness of individuals—their enthusiasm about the training was elevated soon after the training and they expected that such training would improve safety. As time progressed, the participants' enthusiasm declined, and in some cases it even reversed to strong negative feelings, due to 'lack of management follow-up' (Taylor & Robertson, 1995; Taylor, 1998; Taylor, 2000). As we continue to contemplate the meaning of 'lack of management follow-up', we are realizing that perhaps it means different things to different people, and also, at different times. For example, in some cases it might mean that the workforce simply expected a follow-up or a recurrent training program; in other cases, it might mean that the workforce expected the management to solve all the problems identified by the workforce as a result of their new-found knowledge about maintenance human factors; and in some other cases, it might mean that the workers are ready to make changes in their individual work habits, but they also need to see that the management is willing to support organizational changes (examples include scheduled shift overlap to allow for better turnovers, streamlined document change process to allow for easy revisions to maintenance procedures, provision of proper tools to minimize the use of inappropriate tools, etc.). Therefore, two things are clear: first, management support is paramount, and second, that support is different for different people/organizations and at different times.

Assuming that you have been through an awareness-level course in maintenance human factors, we will now focus on specific implementation issues at individual, organizational, and collegiate levels.

Incidentally, if you have not taken a basic Human Factors in Maintenance course, please visit the following websites for a primer:

1. http://hfskyway.faa.gov
2. http://www.iata.org/ps/training/courses/talt10_04.htm
3. http://www.caa.co.uk/docs/33/CAP716.pdf

Individual Perspectives

As a result of awareness-level training programs, we discovered that the majority of the trainees were able to make some passive, individual changes. For example, they said that they were more aware of safety issues, they tried to listen to turnovers more carefully, they double-checked their work, or they understood how stress, fatigue, and distraction could affect the quality of their work. Overall, most people admitted to making some changes that were within their personal span of control.

In order to convert the above passive changes into active changes, we offer guidance in the following skills:

1. *Communication*: Practice closed-loop communication—try to provide feedback to others that you understood their message.
2. *Assertiveness*: Assertiveness includes speech as well as actions. Speak-up to identify discrepancies, but also act to follow through on your responsibilities.

3. *Preparation*: Be prepared, physically, intellectually, and emotionally, to handle the job at hand. If not, alert your colleagues of your weaknesses.
4. *Work Management*: Understand your strengths and weaknesses prior to embarking on the job so that you are able to better manage the tasks at hand: organize, prioritize, and maintain situational awareness. Above all, shield against distractions.
5. *Integrity*: Learn to stand-up to your convictions. Safety is your fundamental responsibility. Do not sign-off all items in one sitting, do not assume anything, and do not forget to document all your actions.
6. *Teamwork*: Teamwork is difficult to practice in the maintenance environment because the reward and penalty structure is based on individual-level certification authority. However, good teamwork can improve safety and job satisfaction for all. Look for opportunities to share your abilities and skills with others on your crew.

Now, let us discuss each of the above areas in more detail using specific examples from the NASA Aviation Safety Reporting System (ASRS) maintenance dataset—a set of de-identified error reports voluntarily submitted by mechanics in the United States.

Communication

Communication involves a transmitter (the one sending the message), a medium (such as technical/legal document or a shift turnover/handover), and a receiver (the one receiving the message). Typically, the transmitter simply sends the message without confirming that the message has been received and understood by the receiver. Since humans are not perfect, sometimes we simply *think* that we have sent the message! Therefore, to *close the loop,* a *feedback system* needs to be implemented. This feedback system should not only confirm that the message has been received, but also confirm that it has been interpreted in the intended sense. Sometimes, the medium used to convey the message tends to impede clear interpretation. Consider Example 1.1. It illustrates how the incomplete and confusing maintenance instructions, impractical tools/equipment, and conflicting organizational policies could setup one mechanic into committing errors. A closed-loop communication process could be used to seek clarification regarding maintenance instructions or to change them so that they can be accomplished as published.

Example 1.1: A case of incomplete and confusing instructions

ASRS Report Number 459526

(Public document, edited for clarity)

> Aircraft xyz returned to zzz with smoke in cabin. Secured leaking APU and performed the required operation xwx. This operation is new at air carrier as of dec/xa/99. The job card refers to maintenance manual 21-20-02-2 for a duct burnout. Found maintenance manual to be very hard to follow as to which sense lines must be capped because instructions do not match illustrations. The instructions call for 2 lines to be capped, but illustrations show 3 lines to be capped. Zzz does not allow for the APU compartment to be pressure washed on the ramp, so compartment was wiped with solvent and rags. This is allowed per job card, but all the oil cannot be removed from acoustic liner of APU inlet. In hindsight, APU should have been placed on maintenance callout. Job card signoff card should have separate signoff for each job function, but several functions are included under 'perform duct burnout'. The job card instructions (step 4) say to do duct burnout per maintenance manual 21-20-02-2. The maintenance manual 21-20-02-2, item 7-c does not tell you to replace the coalescer bags. But if you read job card step 4 instructions, replacing coalescer bags and cleaning the water separators is included under 'perform duct burnout'. The job cards should have a separate signoff for coalescer bag replacement. The tooling provided by air carrier will not allow the burnout procedure to be performed as per maintenance manual. The leads and hoses for test equipment will not reach the cabin as required. Air carrier provided the kit, but did not provide any training on its use or on how to perform the test. Due to the discrepancies in the job card and maintenance manual, as well as the incomplete and vague instructions, I am not completely sure that I performed this operation properly.

Another problem with the most technical publications used to communicate maintenance instructions is that there is no clear way to determine whether the specific instructions that are used by the mechanics are in fact the most current; this is especially true for older aircraft and in remote maintenance facilities.

Interpersonal communication is absolutely critical among aircraft maintenance personnel for the following reasons: (a) it minimizes the probability of injuring oneself or others working on the aircraft, (b) it keeps all mechanics fully informed about the several maintenance actions that tend to be carried out simultaneously and have a tendency of having inter-connected consequences, and (c) it minimizes complacency among people who have worked together for long periods of time. While working in shifts, it is imperative that mechanics from two linked shifts have a face-to-face conversation about the job in progress. In the United States, the typical organizational structure in most airline maintenance organizations is as follows: three-to-five mechanics holding an FAA Aircraft Mechanic Certificate are supervised by a lead mechanic and several leads are supervised by either a foreman or a team leader. Typically, mechanics provide status reports to their leads and the leads provide status reports to the foremen. The outgoing foremen may talk with the incoming foremen, and the outgoing leads may talk with the incoming leads, but the outgoing mechanics rarely talk with the incoming mechanics. Therefore, the actual mechanics taking over the job in progress may have very limited, and sometimes incorrect, information about the job. In the case described in Example 1.2, it is clear that the mechanics were using a 'generic' maintenance manual. Both shifts were aware that the manual was generic. Therefore, the shift that removed the engine and the propeller should have noted any inconsistencies between the pre-existing assembly and the diagram provided in the manual. At least, since most of the people at that maintenance facility worked on 100-series aircraft, they should have noted that the engine control cables were installed on the opposite side on the 200-series aircraft. If they had observed this difference, it would have been prudent to provide a detailed turnover/handover to the incoming shift. It is also interesting to note that the post-rigging inspection was carried out 'in accordance with the maintenance manual'. So, if the maintenance manual was vague or incorrect, the inspection is not likely to be effective in detecting errors.

Example 1.2: A case of an incomplete communication

ASRS Report Number 365609

(Public document, edited for clarity)

On the weekend of April/xx/97, one of our 200 series Dash 8s ferried to our base for an engine change, propeller change, and sheet metal repairs. Our base here in zzz normally only works on 100 series Dash 8s. On Saturday, a crew removed and replaced the P&W 123 engine but did not install the propeller nor did they rig the engine controls. On Sunday another mechanic and I were told by our supervisor to install the propeller and to get the aircraft ready for an xx00 departure. The propeller installation and engine rigging was complied with and inspected in accordance with the maintenance manuals.

The event was caused by the nacelle mounted control rods being installed on the wrong side of the levers at the engine's fuel control. The correct side was opposite of what was in the manual.

> The main contributing factors were that the manuals were incorrect and not specific [to the model being serviced], lack of training, and a poor turnover from the crew that removed and replaced the engine. The situation was discovered during a revenue flight on the next day when the crew reported that the power and condition levers for that engine bound up. The crew was able to free the levers and continued and completed their flight.

Ideally, shift turnovers should include a written narrative, a verbal explanation, and a face-to-face show-and-tell session—mechanics to mechanics, leads to leads, and foremen to foremen. In case face-to-face meetings between mechanics are not practical, the written turnovers/handovers serve as descriptive documentation between the direct parties involved in the performance of the maintenance actions. In one company, we have noticed foremen using turnover logs to communicate with each other, but sometimes, they do not walk over to the aircraft and review the physical status of the job. At least the written logs serve as reminders, and because they are not required by the Federal Aviation Regulations, there is little reason to expect such turnovers other than as a 'best practice' at this particular company.

The U.K. CAA's *CAP 716* (CAA, 2003) presents detailed guidance on shift-turnovers/handovers. We certainly support all those recommendations. To provide further guidance in improving shift-turnovers, we recommend that a 3-copy maintenance log form be used: one copy would stay with the mechanic who gave the turnover; one copy would be given to the mechanic who accepted the turnover; and one copy would be given to the supervisor—via the lead mechanic—to be filed with aircraft records that are to be maintained with the aircraft. The mechanics should retain their copies for future reference. Such a maintenance log would not only assist the mechanics in better communicating the job status, but it is also likely to provide some legal protection to the mechanics in the case that the management releases an unairworthy aircraft into service (see Example 1.3). Furthermore, novice mechanics can use such logs as means to document their troubleshooting skills. As the logs grow, the mechanics will be able to refer back to previous work and improve upon their troubleshooting strategies. A sample illustration of such a form is shown in Figure 1.1. Needless to say that for such a strategy to be successful, the regulators need to view these logs as status reports, similar to pilot logs.

Example 1.3: A case of management releasing an unairworthy aircraft into service

ASRS Report Number 420323

(Public document, edited for clarity)

A B767 was returned to service with a #2 engine oil leak that required an engine change to correct. Report was signed off illegally by a supervisor using an unauthorized signature and employee number.

On the night of NOV/XA/98, my crew worked aircraft for an oil leak #2 engine at station ZZZ. The mechanic that troubleshot the problem was Mr. X. The right engine had a write-up for oil leaking from the compressor rear frame oil scavenge B sump fitting calling for the gasket to be replaced. This was not the problem. The problem was much deeper than first thought. We got Technical Services involved. Told them what we found. Technical Services said that we had an engine change in area of the leak with oil in the tailpipe point to the recoup duct. But he wanted to discuss with an engineer, Mr. A. The supervisor Mr. Z was aware at all times what was going on. Mr. Z knew we had to ferry to station ABC for engine change. At around xa30 I made a logbook maintenance entry stating that the engine had to be changed. When Mr. X and I left work that morning, the original write-up and the engine change entries were still open with the supervisor, technical services, engineer and management (Mr. B) aware [that] the engine would be changed. This did not happen. Mr. Z signed off the original write-up using Mr. X's name and employee number and Mr. B signed the engine change item. Neither Mr. X nor I were ever aware that Mr. Z was going to do some thing like this. The aircraft had one more logbook entry regarding an oil leak #2 before it even left station ZZZ. This engine should have been changed and supervision should have never used an employee's name and number under false pretenses.

Always Safe Airlines Aircraft Type: ______________ Start of Job Date: __________ Registration Number: ________ Start of Job Shift: __________ Job Card Ref. # __________ Outgoing Mechanic/AME's Name: ________________________ Incoming Mechanic/AME's Name: ________________________ Outgoing Lead/Supervisor's Name: ________________________ Incoming Lead/Supervisor's Name: ________________________		
Description of Tasks Accomplished	Initials of Outgoing Mech/AME	Initials of Incoming Mech/AME
Contact Information: Outgoing Mechanic/AME's Phone Number: _______________ Outgoing Lead/Supervisor's Phone Number: _________________	Initials of Outgoing Lead/Supervisor	Initials of Incoming Lead/Supervisor

Figure 1.1: A sample shift-turnover log form

Assertiveness

Assertiveness can be practiced at three levels: first, it refers to speaking-up when one notices a problem or a discrepancy; second, it refers to reporting (typically via an ASRS report or in some cases self-disclosure to the company); and third, it refers to following-up on the root causes associated with factors leading to the discrepancies. Each of the above levels of assertiveness has its own set of obstacles. For example, people may not speak-up because they did not realize that there was a problem—some mechanics (variously referred to as Aircraft Maintenance Technicians [AMTs] or Aircraft Maintenance Engineers [AMEs]) continue to assume that the already installed part was the correct part (see Example 1.4) or a similar repair on another aircraft was correctly performed (see Example 1.5).

Example 1.4: Assuming that already installed parts are correct

ASRS Report Number 461689

(Public document, edited for clarity)

> On Jan/xa/00 an aircraft came in with a logbook entry that the #2 rudder hydraulic caution light came on and stayed on inflight. The lead mechanic told me what part needed to be changed and had gotten the part from the parts room. Next, he showed me where to put the part in the aircraft, which was in the tailcone (the tailcone in the DHC-8 is a poorly lit area with the exhaust pipe from the heat exchanger running through it). After I started to install the first switch (holding a flashlight in one hand and removing and replacing with the other hand) the lead mechanic found out that both switches would have to be replaced. He advised me that I would have to change both pressure switches. The lead mechanic then asked another mechanic to get another switch and to use the part number from the box which the part I was installing came in. Normally, when tasked with a job, I go to the Illustrated Parts Catalog and look up the parts myself, but since he had looked up the information already, I thought he had the correct part numbers. The other mechanic came back with the part and I installed it. I did not know at the time that the second switch I installed should have been a different part number and since the parts look alike, I did not notice the difference in the poorly lit tail cone. The lead mechanic came to me after I had completed installing both switches and closing the access panel to advise me that the switch that needed to be changed was on the rudder actuator and the switches I had replaced were to the powered flight control system. I then went and looked up the part number for the pressure switch on the rudder actuator that needed to be changed and installed it. I then did an operational check, which was good.

> When the aircraft was taxied to the gate, I noticed that the rudder pressure light came on after a delay of about 5-7 seconds. When I did the operational check, I did not wait that long before moving to the next step and it was not noted on the work card. The rudder pressure light was MEL'ed (deferred for later maintenance) and 3 days later was ferried to a maintenance base to be fixed. That's when it was discovered that the second switch I installed was incorrect part. The correct part was then installed.

Example 1.5: Assuming that previous repair was correct

NTSB Case Number AAR 95-04, Learjet 35A

> On December 14, 1994, a Phoenix Air Learjet 35 crashed in Fresno, California. Among the probable causes listed in the NTSB report was an improperly installed electrical wiring for special missions operations that led to an in-flight fire that caused damage to airplane systems and structure. The chain of events leading up to the accident indicates that a series of Special Mission Power Wiring installations were initiated in 1989. For the first installation, an FAA Form 337 was used to document the installation and seek its approval. Subsequently, this repair (without physically reviewing the data that were used to approve the 337) was used to approve similar repair on 14 other aircraft. Unfortunately, the repair that was carried out on the first aircraft was inconsistent with the data in the FAA Form 337. Since nobody else validated the consistency between the FAA Form 337 and the actual repair, they simply duplicated the original mistake...14 times.

Patankar & Taylor (2004, pp. 182-184)

The second obstacle to assertiveness is the fear of persecution. We know from previous research (Patankar, Taylor, & Goglia, 2002) that as many as 30% of the mechanics working for large airlines in the United States do not trust that their managers will act in the interest of safety. Consequently, there is reason to believe that the mechanics may feel uncomfortable reporting safety problems. Nonetheless, mechanics seem to feel comfortable reporting personal mistakes as well as systemic problems confidentially via the ASRS forms. Patankar and Taylor (2001b) reported a steady increase in the number of ASRS reports submitted by mechanics as well as the use of human factors terms such as 'lack of resources', 'pressure', 'distraction', etc. in such reports in the three years following the introduction of ASRS forms for mechanics in January 1996. Therefore, there is some empirical evidence that the assertiveness among mechanics is increasing.

In a large airline or repair station, the mechanics may perceive themselves as ineffective in driving a procedural change to the point where causal factors contributing to the systemic errors are eliminated. So far, the ASRS reports indicate that mechanics are comfortable reporting systemic problems and their challenges in getting organizational procedures changed; however, there are no reports of having been able to actually effect an elimination of causal factors.

As illustrated in Example 1.6, maintenance documents can contain systemic errors that repeatedly set mechanics or inspectors for undue tasks/responsibilities. Only when the end users practice assertiveness by striving for appropriate changes in such erroneous documentation, will the system have any hopes for a change.

Example 1.6: Example of a systemic problem with maintenance documents

ASRS Report Number 366612

(Public document, edited for clarity)

While working as an aircraft inspector for a Part 121 airline, xyz, on April/xx/97, I was required to complete step 6 of job card xxx. This step requires completion of an item condition report and a corrosion prevention and control inspection report. By signing for step 6, as an inspector, I am being forced to attest that the corrosion prevention and control inspection report has been completed. The fact is the corrosion prevention and control inspection report is not prepared by a floor level inspector even though by signing the step the report appears to have been prepared by me. At this time, I have submitted a maintenance program revision request to correct this problem. This is indicative of the lax preparation of the maintenance paperwork and forms that xyz uses to conduct its operations under 14CFR Part 121.

Callback conversation with reporter revealed the following information: the reporter states that when you sign off the work card you sign for accomplishment of the corrosion inspection and the report. This prevention and control inspection report is done by management and is not available to the inspector using the work card.

In another example of assertiveness (see Example 1.7), a mechanic sought authorization from his duty manager to cannibalize a switch panel in order to release a Fokker 100 aircraft. Later, he realized that the cannibalized part was wrong. Fortunately, he was able to detect his error and take appropriate corrective action. This is a great example of how assertiveness should be practiced. There is not much that one could have done to prevent this particular error; however, the consequence of that error could be contained and future similar errors could be avoided if one were to double-check with the maintenance manual. After all, it is the mechanic's signature on the release, not the maintenance controller's.

Example 1.7: Controlling consequences of errors through assertiveness

ASRS Report Number 528060

(Public document, edited for clarity)

> A Fokker 100 was dispatched in non-compliance with the incorrect voice recorder and flight data recorder control panel installed.
>
> On OCT/SUN/01 at about XA00, advised by crew of aircraft X Fokker F100 at ZZZ CVR/FDR test switch inoperative. After calling Maintenance Control Foreman, I was told I would have to replace switch panel before dispatch. None was available at ZZZ. I contacted Duty Manager for rob panel authorization. Duty Manager requested a rob from aircraft Y at hangar and received authorization number from Maintenance Operations Control. I removed panel from aircraft Y and installed it on aircraft Z. A test of switches on panel was normal and logbook signed and aircraft dispatched. After receiving replacement panel and attempting to do data entry in maintenance computer, I discovered that the wrong panel had been robbed and installed instead. Also on looking at maintenance test procedures in maintenance manual Chapter 34, I discovered I had not done a full test of avionics switch panel. I immediately advised Maintenance Operations Control Foreman of my errors and asked him to locate aircraft and correct my errors. I then ordered panel for aircraft Y and sent serviceable part to ZZZ for installation on aircraft Z.

A slight variation of the above issue is that sometimes, Maintenance Control may say that it is okay to release the aircraft, but fail to tell the mechanic that the aircraft should be downgraded. See Example 1.8 for an example of such a situation.

Example 1.8: A mechanic failed to downgrade an aircraft post MEL clearance

Excerpt from a confidential document

> On March/xx/99, an AMT received an air carrier aircraft with an autopilot discrepancy. Since he was not an avionics technician, he contacted avionics maintenance control for guidance. After discussing the problem with maintenance control, they advised the AMT to replace the #1 Air Data Computer (ADC). The AMT accomplished all the necessary steps in replacing the ADC and notified the maintenance control of his findings. They advised the AMT that all his work was satisfactory and that he should clear the item in the maintenance log. The maintenance control failed to advise the AMT to remove the aircraft from CAT-II status and since the AMT did not have much avionics experience, he did not know about this requirement.

Preparation

Preparation involves assessment of the physical, intellectual, and emotional capabilities of the mechanic performing a task. Considering that maintenance workers tend to work in shifts, they are most likely to be out-of-synch with rest of the world. When a night-shift worker comes home, he needs to sleep while the rest of the world is active and making noises. Similarly, the afternoon- or swing-shift workers and the day-shift workers have to adjust their sleeping routines. Their families have to adjust their own expectations as well as routines. Therefore, every time a mechanic reports to work, he needs to evaluate his physical status—has he had enough rest? Motor skills, decision-making abilities, and emotional strength to handle multiple tasks and priorities are likely to be affected by lack of sleep/fatigue. In fact, Dawson and Reid's (1997) study on fatigue equates 24 hours of sleep deprivation to 0.10% blood alcohol content—illegal to drive in most western industrialized countries.

At one maintenance facility, we informally polled the maintenance workers to find out that the majority of them were commuting from more than 50 miles (one way), were working multiple jobs, and were working 4-day, 10-hour shifts. They also admitted to walking around like 'zombies' on the third and fourth days! This information is not just alarming, but it is a fact. We need to find ways to address such challenges.

Examples 1.9, 1.10, and 1.11 illustrate examples of situations needing better preparation.

Example 1.9: Performance impaired due to physical fatigue

ASRS Report Number 448357

(Public document, edited for clarity)

On Sep/xa/99, air carrier airlines flight xyz departed zzz at xa:59. After feeling a vibration in the landing gear on taxi out, the crew returned to the gate. Investigation revealed #1 main wheel bearings had failed. A plane change was made and the flight to abc departed 2 hrs and 6 min late. The aircraft (a B737-300 air carrier plane xxx) was taken to the hangar for further evaluation. A more detailed investigation revealed that the axle was damaged and was replaced along with both brakes, tires and wheel assemblies. The cause of the bearing failure was the inner bearing was the wrong bearing. The tire and wheel assembly with the wrong bearing was installed on Sep/xb/99 and had made 6 landings before the bearings had failed.

I was assigned to the air carrier airlines tire shop in zzz, on aug/xa/99. One of my responsibilities that day was to be the second set of eyes in the installation of the wheel bearing. One man draws the new bearings from the bin, checks part numbers, inspects the bearings then packs the bearings with grease. After the tire assemblies are built up, he installs the bearings and signs off the bearing box on the serviceable tire tag. The second set of eyes was developed to prevent the wrong bearings from being installed. I was that second set of eyes and signed the box designated for that task on the tire tag. One of the B737-300 main wheel and tire assemblies built that day made it to aircraft xxx with the wrong inner bearing. I was responsible for checking the bearing part numbers, and somehow did not get the job done.

I will not try to make excuses but would like to give my thoughts on what might have been some contributing factors in this incident:

1. The bearing that was installed was part number 594 a B757 main inboard bearing. The bearing that was the correct bearing was part number 596. Again, I am not trying to make excuses but the part numbers are very similar and the numbers could get transposed. Also, the sizes of the two bearings in question are almost identical.

2. The zzz tire shop is a very busy place. A crew of 8 AMTs in the summer months build approximately 500 tires a month. In the last six months we have averaged 18.3 tire and wheel assemblies a day. The day of Aug/xb/99 was especially heavy. 34 tire and wheel assemblies were built that day (more tires than any other day in zzz tire shop history) including 1 B757 main, 6 B727

> main, 2 B737-200 mains, 3 B737-300 mains, 16 B737-300 noses, 4 A320 mains, 1 B777-a main, and 1 B747-400 main. The installations of the bearings are done at the end of the day. I think fatigue was a factor.

To prevent future occurrences I am building a Go/No-go gauge that will be a positive check of the two bearings that are so similar in size and part number. The gauge can be used after the bearings and retainers are in place. If approved by engineering, this will be a final check before the bearing shipping covers go in place. This scenario is a problem that has been going on for years. The bearing manufacture (Timken) offered two years ago to color-code the bearing cages, but the company failed to adopt that change.

Callback conversation with reporter revealed the following information: the reporter stated that 34 tires had been built up on this day and only 33 bearing sets were available and one set was ordered for the remaining tire. The reporter said 33 serviceable tires were dispersed to 4 terminal locations for ready access by the technicians. The reporter said he works in the tire build up shop and three possible scenarios can allow the wrong bearing to be installed. The reporter stated the wrong bearing could have been installed in the buildup shop or could have been robbed from a serviceable tire in the gate storage area for a bearing dropped in the dirt by a mechanic. The reporter said the third possibility is the manufacturer boxing the bearings with the wrong identification and the shop not checking the bearing part number and installing the bearing inadvertently. The reporter stated the real problem is a [need for] fool proof bearing identification method. The reporter said he submitted a Go/No-go gauge which checks the slight difference in bearings. The reporter said the company engineering department rejected this suggestion. The reporter stated the bearing manufacturer has suggested color coding the bearings but this idea was not acted upon.

Example 1.10: Performance impaired due to lack of knowledge of the task

ASRS Report Number 395238

(Public document, edited for clarity)

During an FAA inspection, inspectors found insufficient training on the deHavilland -8 in the composite shop. FAA inspectors found that repair procedures may have been deviated from the correct procedures. During the time that I have spent in the composite shop, I may not have complied with repair procedures completely due to lack of composite training.

The company was made aware that it had not provided training in composite repair to specific aircraft type.

Callback conversation with reporter revealed the following information: the reporter stated that prior to the FAA shop inspection very few people were trained on the repair and buildup of composite structural components. The reporter said that some of the work done by untrained mechanics was not per the manufacturer's structural repair manual and was never inspected. The reporter stated that after the FAA shop inspection, the company started a training program and now maintains training records of mechanic qualification.

Callback conversation with reporter of ASRS Report Number 394876 revealed the following information: the reporter said the shop operated with production as the primary goal with repairs not being done per the structural repair manual which cut job time and increased production. The reporter stated there never was a training department and no training records were maintained in the past. The reporter said that prior to the past few months, a lot of unapproved repairs went out of this shop and were never inspected.

Example 1.11: Performance impaired due to emotional distractions

ASRS Report Number 401533

(Public document, edited for clarity)

On May/xy/98, I was given the work cards to close up the left wing of aircraft xyz in accordance with the company's 'C' check program. After installing the panels and stamping off that all panels had been installed, I later found out that 2 panels on the #1 leading edge slat were not installed on the aircraft prior to flight. I did not observe that any panels had been removed from the #1 slat during the period in which I was installing the leading edge panels. I have also found that there has been a revision in the paperwork recently to remove the panels on the #1 slat for inspection. I feel this incident could have been avoided if the out going inspection paperwork were to read, 'Inspect aircraft with slats and flaps in the full down position'. Therefore, any missing panels could be seen at that time prior to dispatch. Currently, we are trying to get the inspection paperwork changed so this will not happen again.

Our company for the past year has been telling us they are going to relocate our work and close three maintenance facilities. I feel that this has caused some distractions from us doing our jobs efficiently. The uncertainty of not knowing where we will go, having to sell our homes, move, and buy new homes has had a large effect on everyone.

Work Management

Work Management involves organization, prioritization, and execution of one's work. In a base maintenance situation, you may have more time and flexibility to achieve better work management; however, in line maintenance you just might have to depend more heavily on others. For example, if you are working in base or major maintenance, you are likely to perform a rather involved maintenance task. When you are faced with the need for new hardware or part, do you use the old hardware/part as a reference or do you actually refer to the Illustrated Parts Catalog or maintenance manual? We have already presented you with some cases (Examples 1.4 and 1.5) to illustrate how mechanics tend to use previously installed parts or previously conducted repairs as a reference rather than proper technical documents. From the perspective of Work Management, we would like to stress the importance of recognizing such hazards early and planning to circumvent them.

The two key issues under Work Management are as follows:

1. *Strategy to manage variance*: aviation maintenance is a very dynamic and highly unpredictable environment. The maintenance instructions, job cards, federal regulations, etc. are based on certain assumptions of this work environment and the qualification/experience of the people performing the maintenance tasks—devoid of challenges such as time pressure, lack of resources or training, the company's financial status, etc. Therefore, there is a certain degree of variance between what is expected by policies, procedures, and regulations versus what actually happens in the real world.

   We will explore the issue of variance from a sociotechnical perspective in Chapter 2, but it is important to note that variations from normal expected conditions tend to increase risk. For example, if the maintenance manual expects that a field maintenance facility will have equipment to test avionics units such as the Distance Measuring Equipment (DME), but you find out that your facility does not. Now, you are tasked with the challenge of managing this variance. Similarly, suppose an engine change is expected to take 6-8 hours, but in your particular case, for whatever reasons, the change has gone beyond 10 hours. Now, you are tasked with the challenge of managing variance associated with time pressure.

2. *Strategy to manage in the void*: This is another sociotechnical concept that we will discuss in greater detail in Chapter 2, but for now let us

examine its context. As you are well aware, most maintenance tasks have specific procedure to guide the mechanic; however there are some tasks for which there are no published procedures. Typically, such tasks are infrequent and the mechanic is expected to improvise. *Management in the Void* refers to the intuitive framework used by the mechanic in making decisions under conditions of such uncertainty. From the perspective of Work Management, the mechanic is expected to be able to evaluate the risks, make decisions, evaluate the decisions, and apply the necessary corrective actions.

Integrity

First, aviation maintenance professionals are among the most ethical and law-abiding individuals because their entire professional training is based on regulatory compliance. In recent times, however, we have discovered a few instances of ethical creep, a normalization of deviance of sorts. Ethical creep or deterioration of integrity surfaces in the following ways: (a) mechanics releasing aircraft when it is 'almost within limits', (b) lead mechanics/managers assigning mechanics to work on an aircraft when they know that the mechanics are not adequately qualified, (c) supervisors stopping the inspectors from inspecting because they cannot afford any more write-ups, (d) supervisors signing-off aircraft that their mechanics refused to release, and (e) mechanics signing-off all the items on a job card in one sitting, sometimes inadvertently signing for tasks that they did not accomplish. Of course such deviations are not so frequent that aircraft are falling out of the sky every week—not yet. However, we are compromising the design safety of the aircraft, and the public's confidence in our ability to provide a safe aircraft. As professionals, it is important for us to recognize these deviations in integrity and make every attempt to improve our behavior.

Example 1.12: One mechanic released the aircraft because it was 'almost within limits'

ASRS Report Number 48638

(Public document, edited for clarity)

On Sep/xa/00 at approximately xa15 while closing in some safety sliders on our aircraft docking structure, I accidentally created damage to the skin of ship xyz of air carrier B757 fleet with the tool we use to move the safety sliders. The aircraft was in its last couple of days of overhaul. I do not remember the exact location of the damage but I believe it was located

> somewhere between station 440 and station 540 about 10-25 inches above the cabin floor line on the left-hand side. I reported the damage and a non-routine work card was generated. The damage was initially evaluated as a dent and was to be checked to see if it was in limits per B757 Structural Repair Manual 53-00-01-1-101. The dent was smooth but irregular in shape except for the very center where it appeared to have a small amount of paint which had been pushed into a little mound so to speak. Examination revealed no bare metal gouges so it was concluded that it was in fact a dent. It was determined that the dent was technically within limits when using the edges of the outer indentation. However, if you look at the most severe area of the indentation itself, it may not be. A couple of days later, I began to doubt if the measurements I used were actually the ones I should have used. The minor axis dimension I used was the smallest of the outer indentation, but I wonder if the minor axis dimension should actually be from the edges of the inner indentation. I am not an engineer, but the more I think about it, it seems to me that if there is any area of the dent that is out of limits, then the whole thing should be. This possibility is not addressed in the SRM reference we used. Inspection bought it and the aircraft is now in service.

Deciding whether or not a particular damage or wear is within limits is a critical challenge in maintenance. Even if one were to make a purely scientific measurement, the results are likely to vary simply due to variations in the way the measurements are taken or due to instrument errors. However, if we were to also consider the operational factors such as risks involved in declaring a false negative (declaring that a repair or wear is not beyond limits when in fact it is) and the perception of such a risk to the safety of flight is low, it is likely that the individual will be predisposed to declaring a false negative.

Example 1.13: One lead mechanic assigned non-ETOPS qualified mechanics to work on ETOPS aircraft

ASRS Report Number 457371

(Public document, edited for clarity)

> I am an A&P mechanic for a major airline. On nov/fri/99 a junior mechanic and I were assigned 2 tasks on a B767-300 ETOPS aircraft by our lead mechanic in charge of the shift. At the time the tasks were assigned, I was classified as a junior mechanic. The tasks were to remove and replace an APU and to install an APU surge valve on the new unit. We completed the replacement of the APU and surge valve per the maintenance manual, performed all operational checks, leak check, and the unit was found to be ok for service. The lead mechanic also performed

an operational check of the unit. I completed the logbook entries for the items I worked and then an ETOPS-qualified mechanic signed the airworthiness release of the aircraft.

On Dec/thu/99 I overheard my foreman discussing with a lead mechanic about a non-ETOPS-qualified mechanic who had replaced an item which required an ETOPS-qualified mechanic to perform the task. I then told my foreman that on Nov/xx/99 I was assigned to perform an APU change and APU surge valve installation on an ETOPS aircraft and that I also was not ETOPS-qualified. I then asked him if an APU replacement and surge valve installation required someone to perform these tasks who was ETOPS-qualified. After some discussion about the matter, we looked in the procedures manual for exactly what tasks were or were not ETOPS items. As it turned out, an APU replacement and surge valve installation are ETOPS items. So I was not in compliance by working on the APU and surge valve installation and not in compliance for the logbook entry. We also found out that the installation of the APU required several job instruction cards to be signed off by the mechanic performing the task and by a mechanic who inspected the installation. On Dec/tue/99 my foreman notified our Company FAA liaison of the situation. The liaison then had an ETOPS qualified mechanic re-inspect the APU and surge valve installation on Dec/yy/99 and he ok'ed the aircraft for ETOPS operations.

Example 1.14: One supervisor stopped an inspector from inspecting because he couldn't afford any more write-ups

Excerpt from a confidential document

An inspector was inspecting the floorboards of a transport category aircraft. He found serious corrosion and several cracks. According to the inspection procedures, he was expected to inspect 18 inches on either side of the corroded area. Since corrosion is not at all acceptable, he continued to peel-off floorboards to inspect them and write them up. Finally, the production supervisor came to the aircraft and stood on the remaining floorboard and instructed the inspector to stop inspecting further. His aircraft was late and he could not afford any further write-ups.

In summary, it seems like the individual integrity in aviation maintenance is challenged by a fundamental role ambiguity: mechanics feel that they are held accountable for safety and managers are held accountable for on-time performance. This would not present such a serious challenge if either (a) safety truly held priority over performance/economics or (b) managers were not allowed to return aircraft to service. However, in the 'real world' these distinctions are not the case and all parties are held accountable for safety, but only management answers up the line for productivity. The role-conflict managers face between joint expectations for high safety and high productivity is often as difficult for them to deal with, as mechanics' facing their role ambiguity just described.

Teamwork

It should be clear to anyone who watches airliners prepare for departure that each flight represents a remarkable degree of teamwork from a very diverse range of individuals and sub-teams. All of them are connected to each other in two aspects: schedule and trust. Everyone on the team—from gate agents to flight crew—has specific roles to play at specific times, and each trusts that the other is doing their job to the best of their abilities. It is an absolutely remarkable symphony!

When we scale this teamwork down to maintenance tasks, we realize that mechanics depend on parts suppliers to provide them with good quality parts, they depend on the company to provide them with accurate and up-to-date maintenance instructions, and they depend on other teams such as Maintenance Control to provide them with reliable guidance on specific Go/No-go decisions. Also, they must depend on each other to honor their professional etiquette.

The potential for strong teamwork is challenged by extremely high responsibility and accountability on the part of the *individual* licensed/certificated engineers/mechanics. As far as the regulatory violations are concerned, it does not matter *why* a particular regulation was violated. The facts are that it was violated and the compliance responsibility rests at the individual level. Therefore, the regulators tend to enforce them at the individual level. Consequently, the individual mechanics are very conscientious, and rightfully so, of their responsibilities. However, from the perspective of teamwork, they have to rely on a wide variety of people who have very little, if any, professional accountability.

Recommendations

Overall recommendations to all maintenance personnel are as follows:

1. *Management personnel*: Do not assign mechanics/engineers to jobs for which they are either not adequately trained or for which they do not have adequate resources such as tools and the latest maintenance instructions. It is your responsibility to ensure that your workforce is adequately trained, equipped, and prepared to do their job in a professional manner.

2. *Mechanics (AMTs and AMEs)*: You are the last link in the maintenance chain.
    a. Never sign-off multiple items on a job card without verifying that each item was in fact accomplished in accordance with the approved maintenance instructions—every time.
    b. Never use old parts as reference to replacement parts; always use the IPC or Maintenance Manual to get the correct part number.
    c. Always check the reference repair schemes for accuracy prior to duplicating.
    d. Never assume a verbal okay from Maintenance Control or another individual without verifying it for yourself via a neutral third-party, even if the verification takes place after releasing the aircraft.
    e. Remember that risk is increased when one intermediate component is disturbed from its original configuration/orientation to gain access to another component.
    f. Lack of time, resources, knowledge, or good overall health increases the risk of committing maintenance errors. When multiple risk-inducing factors are present, the effect is exponentially higher.

3. *Both managers and workers*: Recognize that you are a member of a large, complex team. As such, it is your responsibility to be assertive—both in words as well as in actions—regarding any safety related inconsistencies that you may discover. It is also your responsibility to perform your job to the best of your abilities.

Organizational Perspective

A review of the history of the human factors programs in maintenance indicates that their success tends to be attributed at the individual-level and failures tend to be attributed at the organizational-level. As individuals, most mechanics are more aware of the safety-related effects of human performance issues; as organizations, most organizations seem to be stumped as to what to do next. The next generation Maintenance Resource Management (MRM) programs demand that organizations be willing and able to change their structure as well as processes—provide specific feedback to safety-related recommendations, establish secure and effective self-reporting systems, provide a simplified and effective process to update/change maintenance procedures, reward safety-compliant behaviors, and above all practice higher ethical standards.

Most of the MRM programs in the United States have been aimed at raising the general awareness about human performance factors. Without a specific regulatory mandate such as those imposed by the U.K. Civil Aviation Administration (BCAR Section L, Module 13), Transport Canada [Canadian Aviation Regulation 566.13(b)(iv)], and the International Civil Aviation Organization (ICAO) Annex 6, the U.S. airline industry has been content with the initial awareness-based programs. In recent years, however, some U.S. airlines are gearing-up to offer basic as well as recurrent human factors training because in order for these airlines to be able to do business in Europe, under the Joint Aviation Authorities' (JAA) Regulation §145, they have to comply with the JAR 145 requirements.

Some of the challenges at the organizational level include the need to prove a positive return on investment, the need to reinstate the ethical and professional mandate of 'safety first' among maintenance managers, the need to manage the fine line between reckless behavior and genuine mistake, and the need to establish a secure and effective error-reporting system.

In response to the above challenges, the three major aviation regulatory agencies have reacted in different ways. The European Joint Aviation Authorities have specified requirements at both individual AME level as well as the organizational level. Consequently, individual AMEs holding a JAR-equivalent license will have to demonstrate their knowledge in human factors. Also, aviation organizations, approved as JAR-145 organizations, will have to demonstrate their systemic capabilities of providing recurrent training as well as managing maintenance errors.

Transport Canada has adopted a more systemic solution. They formulated a Safety Management System (SMS) concept under which senior management will be held accountable for the overall management of safety within their organizations. Individual training, reporting of errors, resolution of error-inducing conditions, etc. are all under the SMS umbrella.

The Federal Aviation Administration continues to encourage individuals and organizations to adopt human factors principles. They have released two advisory circulars (AC 120-16D and AC 120-79) and one *Maintenance Resource Management Handbook* to provide general information about maintenance resource management programs, guidance regarding the management of self-reporting systems, and guidance regarding the more systemic Continuing Analysis and Surveillance System (CASS). Overall, aviation organizations in the United States have ample resources to develop, implement, and manage a successful error management program in aviation maintenance. Furthermore, U.S. companies intending to do business in Europe will have to comply with JAR 145 requirements.

The Continuing Analysis and Surveillance System (CASS)

In April 2003, the Federal Aviation Administration released an Advisory Circular (AC 120-79) delineating their expectations regarding the development and implementation of a CASS program. Apparently, the regulatory requirement for such a program has been in place for several decades. The actual requirement is as follows:

> Each certificate holder shall establish and maintain a system for the continuing analysis and surveillance of the performance and effectiveness of its inspection program and the program covering other maintenance, preventive maintenance, and alterations and for the correction of any deficiency in those programs, regardless of whether those programs are carried out by the certificate holder or by another person.
>
> (14CFR §121.373(a))

In a letter to the U.S. Department of Transportation Office of the Inspector General, Frank Boksanske of the Aircraft Mechanic Fraternal Association (AMFA), a labor union representing over 12,000 aircraft mechanics in the United States, claims 'Even though CASS has merit, the program is left incomplete, subject to erroneous data collection, with final analysis left to the carrier itself' (Boksanske, 2002). One major reason for the lack of confidence in the CASS program is that it was developed without considering input from aircraft mechanics actively involved in maintenance. Although this claim by AMFA has some merit, the AC 120-79, published subsequently, seems to provide enough details regarding the expectations of mechanic involvement and the role of management. It is hoped that this advisory circular will help air carriers in developing and implementing a successful CASS program.

From the perspectives of safety and human factors, the CASS program is intended to proactively solve systemic problems through rigorous data analyses. Data from maintenance write-ups, manual corrections, parts orders, event investigations, flight delays due to maintenance, quality audits, etc. are to be analyzed by specialists who are familiar with the maintenance processes at the particular air carrier. With such detailed guidance in place, it seems like the FAA field inspectors have specific criteria with which to evaluate the compliance with 14CFR §121.373(a) more strictly as well as consistently. Considering the heavy workload of the FAA field inspectors coupled with the fluctuating strength of the airline industry's financial status, whether those inspectors will actually hold the airlines to the CASS requirements remains to be seen. Nonetheless, regulatory basis as well as the requisite guidance material is now available for air carriers to develop and implement organizational changes so that the quality (and associated safety) of maintenance can be improved.

The JAR 145 Requirements and the U.K. CAA's Recommendations

In December 2002, the U.K. CAA published a document, *CAP 716*, providing recommendations to aviation maintenance organizations on how they may satisfy the JAR 145 requirements. The JAR 145 requirements specify the following ten topics:

1. General/Introduction to Human Factors
2. Safety Culture/Organizational Factors
3. Human Error
4. Human Performance and Limitations
5. Environment
6. Procedures, Information, Tools, and Practices
7. Communication
8. Teamwork
9. Professionalism and Integrity
10. Organization's Human Factors Program

The U.K. CAA's guidance materials regarding the above requirements are consistent, in spirit, with the FAA's CASS guidance. The *CAP 716* addresses a broad range of topics including basic training, safety culture development, error reporting, management and tracking of document changes, shift handover protocol, etc. The implications of the JAR 145 requirements are particularly significant to the U.S. airlines conducting business in Europe because many of the U.S. airlines have terminated their maintenance human factors training programs. These airlines will not only have to reinstate their training programs, but will also have to provide organization-level structures and process to comply with both JAR 145 and 14CFR §121.373(a) requirements. Ultimately, both individual-level as well as organization-level changes are inevitable.

Collegiate Perspective

As the industry continues to train its workforce and implement better safety practices, it is essential that the academic community also incorporates appropriate human factors principles in their classrooms and laboratories. Sustenance of changes in safety practices is a cultural change, and as such, it needs to take place at both workforce-level as well as collegiate-level so that the next generation of maintenance professionals will learn safer work habits prior to entering the workforce.

At the time of this writing, none of the aviation maintenance schools at the collegiate level (junior college or university) offers a dedicated course in maintenance resource management or maintenance human factors. Conversely, almost all the flight schools offer courses in general human factors and/or crew resource management. Of course, we can say that the flight students are required, by regulation, to demonstrate their knowledge of human factors while the maintenance students are not. As our previous discussion regarding the CASS program and the JAR 145 requirements indicates, perhaps it is time that the collegiate maintenance programs provide a course in maintenance human factors and/or maintenance resource management.

Considering that most collegiate aviation programs already offer an introductory course in aviation human factors, we recommend an additional course in maintenance resource management containing topics listed previously under the JAR 145 requirements. All the JAR requirements can be taught and practiced effectively in an academic setting. All laboratory courses associated with a typical maintenance curriculum (in accordance with 14CFR §147 requirements) could easily adopt the human factors principles from a practical behavioral perspective. At this time one can only hope that if proper human factors knowledge and skills are provided in the academic environment, the students will develop better awareness and practice of safety-compliant behaviors when they enter the workforce.

Suggested MRM Course Syllabus

We suggest the following syllabus for a course in maintenance resource management.

1. Introduction to Maintenance Resource Management in Aviation
    a. A brief review of maintenance-related accidents
        i. Continental Express EMB 120, Sept. 11, 1991
        ii. Phoenix Air 35A, Dec. 14, 1994
        iii. ValuJet, Flight 597, June 8, 1995

            iv. British Midlands, Feb 23, 1995
    b. Global human factors training initiatives
        i. Federal Aviation Administration
        ii. U.K. Civil Aviation Authority
        iii. Transport Canada
2. Unique Human Performance Factors
    a. Working in shifts
        i. Fixed shifts
        ii. Rotating shifts
    b. Environmental Factors
        i. Poor lighting
        ii. Snow, rain, heat, humidity, etc. affecting human performance
    c. Organizational Factors
        i. Lack of resources
        ii. Lack of training
        iii. Poor Norms
    d. National Cultural Effects
        i. Individualistic or Collectivistic?
        ii. Value/role of assertiveness
    e. Professional Cultural Effects
        i. Can you accept your mistake?
        ii. Will you report your mistake?
        iii. Do you trust your supervisor to make safety-conscious decisions?
3. Individual Coping Strategies
    a. Risk-based evaluation of maintenance tasks
        i. Are you prepared—intellectually, physically, and emotionally to perform the assigned task?
        ii. What strategies are you going to use to minimize the risk of making a mistake today?
        iii. What steps are you going to take to minimize the recurrence of error-inducing situations in the future?
    b. Self-reporting
        i. Does your organization support/encourage self-reports of maintenance errors? Are you comfortable reporting such errors?
        ii. If your organization does not have a reasonable reporting system, do you know of external reporting systems such as the ASRS system—

Aviation Safety Reporting System (U.S.), the CHIRP system—Confidential Human Factors Incident Reporting Programme (U.K.), or the WSDRS–Web Service Difficulty Reporting System (Canada)?

   c. Communication & Teamwork
      i. What specific techniques can you use to improve your interpersonal communication—verbal as well as written, especially during shift turnovers or handovers?
      ii. How can you practice better teamwork? Consider working with other mechanics, leads, external departments, as well as the regulators.

4. Organizational Strategies
   a. Building a safety culture
      i. What is a safety culture?
      ii. What are the elements of a good safety culture?
      iii. How do we build one?
      iv. How do we sustain one?
   b. Quality and Safety Measurement
      i. How do we measure the quality of maintenance?
      ii. What can we do to improve it?
      iii. How do we measure the safety in our organization?
      iv. What can we do to improve it?
   c. Information Management System
      i. How do we collect meaningful data?
      ii. How do we manage, store, and secure these data?
   d. Who is Responsible for Safety?
      i. What are the responsibilities of the senior management?
      ii. What are the responsibilities of the mid-level management?
      iii. What are the responsibilities of the line workers?

Recommended reference materials: We suggest the following reference materials to support the above syllabus:

1. ATA-U.S. Air Transport Association (2001). *Spec 113: Maintenance human factors program guidelines.* Retrieved January 15, 2002, from http://www.airlines.org/public/publications/display1.asp?nid=938.

2. CAA (2002, April). *CAP 712: Safety management systems for commercial air transport operations*. U.K. Civil Aviation Authority, Safety Management Group. Retrieved May 14, 2003, from http://www.caa.co.uk/docs/33/CAP712.pdf.
3. CAA (2002, January). *CAP 715: An introduction to aircraft maintenance engineering human factors for JAR 66*. U.K. Civil Aviation Authority, Safety Regulation Group. Retrieved May 14, 2003, from http://www.caa.co.uk/docs/33/ CAP715.pdf.
4. CAA (2003, December). *CAP 716: Aviation maintenance human factors: Guidance material on the UK CAA interpretation of Part 145 Human Factors and Error Management requirements*. Retrieved on March 20, 2004 from http://www.caa.co.uk/docs/33/CAP716.pdf
5. FAA (1998). *Aviation human factors guide* (v3.0) [CD-ROM]. Washington, DC: Federal Aviation Administration. Available from http://hfskyway.faa.gov.
6. FAA (1999). *FAA Maintenance Resource Management handbook*. Washington, D.C.: Federal Aviation Administration.
7. FAA (2001). Advisory Circular 120-72 Maintenance Resource Management training. Washington, DC: Department of Transportation.
8. Helmreich, R., & Merritt, A. (1998). Culture at work in aviation and medicine: National, organizational, and professional influences. Aldershot, U.K.: Ashgate Publishing Limited.
9. Patankar, M., & Taylor, J. (2004). *Risk management and error reduction in aviation maintenance*. Aldershot, U.K.: Ashgate Publishing Limited.
10. Reason, J. (1997). *Managing the risk of organizational accidents*. Aldershot, U.K.: Ashgate Publishing Limited.
11. Taylor, J.C., & Christensen, T.D. (1998). *Airline Maintenance Resource Management: Improving communication.* Warrendale, PA: Society of Automotive Engineers.

Recommended video tapes: We recommend the following videos to supplement the above syllabus:

1. *Human Factors in Aviation Maintenance*. Video by Transport Canada. Available from http://www.tc.gc.ca/civilaviation/systemsafety/pubs/menu.htm. NOTE: Several other multimedia presentations, videos, and posters are also available from this site.

2. *Not Another Safety Film!* Video by the Maintenance and Ramp Safety Society. Available from http://www.marss.org/videos/video_home.htm. NOTE: Several other multimedia presentations, videos, and posters are also available from this site.
3. *Every Day.* Video by the International Federation of Airworthiness. Available from http://www.ifairworthy.org/ thevideo.htm.

Other Recommended Strategies: We recommend that you consider the following items for inclusion in your course:

1. Role-playing—Simple exercises that illustrate, through student participation, that it is difficult to communicate across professions, especially between pilots and maintainers.
2. Verbal communication game—In this game, a long verbal message is initiated with one student and that student is requested to relay the message to another student. By the time this message gets to the fourth or fifth student, the loss in verbal communication is evident. This game is very effective in illustrating the pitfalls of verbal turnovers/handovers.
3. Written communication game—In this game, students are paired-up in teams of two. On member of the team is asked to write instructions on how to assemble a fairly simple puzzle or model (no diagrams are allowed). The other member of the team is asked to follow the written instructions and assemble the puzzle/model. Again, the pitfalls of poor written communication are evident.
4. Use the *Lost at Sea* teamwork exercise to illustrate how team performance is better than individual performance. This exercise can be purchased from http://www.josseybass.com/WileyCDA/WileyTitle/productCd-088390246X.html.
5. Behavior modeling—ultimately, we have to accept that we must practice what we preach! If we want our students to practice behaviors compliant with their human factors training, we as faculty members must also practice such behaviors. Laboratory environment and university flight line are excellent locations to practice desirable behaviors on a regular basis. Similarly, having posters such as the 'Dirty Dozen' or 'Magnificent Seven' (available from http://www.marss.org) will serve as constant reminders.

Chapter Summary

In this chapter, we emphasized that human factors principles should be applied at individual, organizational, and collegiate levels. At the individual level, we took a risk-management perspective and presented how communication, assertiveness, preparation, work management, integrity, and teamwork could be used to apply human factors principles. At the organizational level, we discussed the impending need to apply the JAR 145 and the 14CFR 121.373 requirements to provide effective structures and processes that facilitate the implementation and evaluation of human factors principles. Lastly, at the collegiate level, we encouraged the collegiate aviation community to develop and implement a maintenance resource management course. The topical outline and the course materials suggested in this chapter should provide ample help to the instructors who want to develop such a course.

Review Questions

1. As a person responsible for the safety and airworthiness of an aircraft, how do you apply human factors principles in your daily life?
2. As a person responsible for the management of maintenance activities, how do you apply human factors principles in your daily life?
3. What specific systemic changes have you implemented/noticed in your organization?
4. Do you think your organization is ready to implement the type and scope of changes recommended by JAR 145 and 14CFR §121.373?
5. What do you see as some of the obstacles in improving the safety and quality of aircraft maintenance in your organization? What can you do about it?
6. As a collegiate faculty member, how do you see human factors principles being implemented in your institution/department?
7. As a maintenance student, what specific improvements would you like to accomplish in the aviation maintenance industry?

Chapter 2

Moving from Awareness to Implementation

Instructional Objectives

Upon completing this chapter, you should be able to accomplish the following:

1. Describe the typical positive and negative effects of awareness-based human factors or maintenance resource management (MRM) training.
2. Explain some strategies that could be used to move toward the implementation of maintenance human factors principles.
3. Discuss sociotechnical systems perspective as it relates to the aviation maintenance environment and to a more action-based MRM program.

Introduction

In our previous volume (Patankar & Taylor, 2004 Chapter 3), we discussed the evolution of Maintenance Resource Management (MRM) programs in terms of four generations. The first three generations of MRM programs focused on raising the awareness of human performance issues as they applied to the aviation maintenance tasks and environment. The fourth generation programs, however, have started to move toward a more direct approach of behavioral change through skills training. In this process, the concept of risk management is gaining some attention because one of the key skills a maintenance professional needs to have is the ability to make decisions based on a thorough consideration of the risks involved.

Risk Management and Safety

Jerome (Jerry) Lederer (b.1903, d.2004) was perhaps the most important single force for aviation safety in the United States. From 1926, until his death, Lederer was actively and successfully engaged in safety improvements for maintenance as well as for air traffic control and the flight deck. Among his accomplishments, he dramatically improved the safety record of the US airmail system in the 1920s, he founded and directed the first US Civil Aviation Administration in the 1930s, he organized the Flight Safety Foundation in the 1940s and ran it for 20 years, and was long an advisor and consultant to the FAA, NASA, the Defense Department and to aviation-related programs in numerous universities. Perhaps the most lasting legacy of Jerry Lederer is his influence on our language and perception of safety. He said,

> Risk management is a more realistic term than safety. It is ever-present and must be identified, analyzed, evaluated and controlled, or rationally accepted. Accepting the premise that no system is ever absolutely risk-free—or conversely, that there are certain risks inherent in every system—it becomes an absolute necessity that management should know, understand and take responsibility for the risks that it is assuming.

Presently, airline management is not well aware of its risks, especially risks taken in repair and maintenance of its aircraft.

Social Role and Human Errors

Frequently, maintenance mistakes and errors are made unwittingly and remain undetected and uncorrected until the next time the aircraft is opened up for maintenance (sometimes months later), or unless trouble is experienced on a later flight. Because mechanics do work alone much of the time, they must be vigilant to catching and correcting their own mistakes—and in real time while they are doing the work. This first type of risk is inherent to the maintenance process. In our previous volume, we called it 'Good Samaritan Risk' (Patankar & Taylor, 2004, p.124). The very fact that a certain maintenance action is being performed creates the risk that there is a probability of an error. The individual mechanic's inability to recognize and correct his/her own mistakes, especially when it is associated with catastrophic consequences, is most often publicized and recognized. Our research on human errors in aviation maintenance over the past decade and a half confirms the usefulness of appealing to individual mechanics to fight their own complacency in reducing errors (Taylor & Patankar, 2001).

Often though, unwitting mistakes and errors are the result of people working together, and of their assumptions and expectations about one another. Social psychologists call these assumptions and expectations the outward manifestations of 'social role', in human interaction. Role is a powerful social mechanism in *causing or correcting* human errors. As you will note throughout this book, real cases from the maintenance archives of NASA's ASRS include many instances where well-meaning mechanics, working under pressure, divide the work such that one removes an offending part while the other orders its replacement. Both are qualified and have probably worked together in the past. The trust and expectation of one another to use correct procedures often cause errors of using incorrect parts or installing them incorrectly. One mechanic will simply not check or confirm all the work done by the other before signing and releasing the aircraft for service. The result is a second type of risk, an unwitting error compounded by the complexities of social expectations and social role. Earlier, we referred to such a risk as 'Normalized Risk' (Patankar & Taylor, 2004, p. 124) because the social roles and expectations are formed over a period of time—initially, mechanics working together may not trust each other as much, but as they get to know each other's work ethic, they tend to become complacent. Management is usually unaware of the size of risk this failure to recognize social complexity represents. Our research over the decade of the 1990s suggests that this risk of errors due to unquestioned social assumptions in mated work is high. For example, in a study of causal factors associated with 939 ASRS reports by mechanics, Patankar (2002) reported that the top two individual-level factors were 'lack of awareness' and 'complacency'. These two factors are associated with two distinct groups: lack of awareness is associated with less experienced mechanics and complacency is associated with more experienced mechanics. Thus we can say that as experience increases, Good Samaritan Risk is likely to decrease and Normalized Risk is likely to increase.

In order to minimize such risks in a typical aviation maintenance facility with its typical maintenance culture, the process of risk management must be viewed as an important, deliberate, and necessary process that is an expected function of maintenance professionals. There is some evidence that success can be gained from training and encouraging managers and mechanics to communicate assertively with one another (Taylor & Christensen, 1998 Chapter 10). Without such special efforts to encourage and support open communication the maintenance culture stifles individual attempts to work with others in risk management. For instance in a typical aviation maintenance facility where formal changes were not successful, if one mechanic took it upon him/herself and insisted on checking the other's work prior to signing the logbook or task card, such a behavior could be seen as a sign of disrespect or distrust of the other—revealing an expectation which would make it difficult for them to work together cooperatively in the future.

Sometimes mechanics and/or managers are aware of their errors, but proceed without correction in the rush and pressure of meeting flight schedules. If individuals speak out for safety and urge correction or delay under these circumstances they may or may not be heeded. In many aviation maintenance facilities assertive individuals can be told, 'That's just the way we do it around here', or 'I am the boss and I say it's OK'. If that kind of situation remains unreported, it remains uncorrected unless and until an incident or accident occurs. Once again airline management is not aware of the size of the risk represented by social pressure exerted by supervisors and other maintenance professionals. This is the third type of risk, the one represented by a conscious error resulting from management and other social pressure in resolving the speed-accuracy trade-off. Previously, we called such risk 'Blatant Risk' (Patankar & Taylor, 2004 p.126).

Risk and Safety-related Information in Aviation Maintenance

This chapter presents how risk management and process control are addressed by individuals, by groups, and by different cultures. The approach taken is to review the 'how', 'what', 'who', and 'why' of risk and safety-related information in aviation maintenance.

How do we Learn about Safety-Related Information and how do we use it?

We often learn about safety-related information in terms of error/performance statistics—a database (often computer-based) of maintenance-caused errors and damage is held within an airline or outside it. In the past, for example, an airline company might keep its own records of aircraft ground damage or of air turnbacks and pilot-reports of maintenance defects, or of inspection department work rejects. Almost without exception these data were not available to maintenance supervisors and mechanics. Also in the recent past, airlines did not keep track of the maintenance labor force's lost time injuries, but would transfer that responsibility to insurance companies providing their Workmen's Compensation coverage, making those data remote and relatively inaccessible to the employees who manage the risks. More systematic efforts by individual airlines take the form of a database created by the Maintenance Error Decision Aid (MEDA) (MEDA, 1994; Allen & Rankin, 1995; Rankin & Allen, 1996). Although many airlines use (or used) MEDA, there are few reported cases of the resulting error database being used for risk management. At best the MEDA program is reported to be used to correct systemic sources of maintenance errors on a case-by-case basis or, and at worst, MEDA data are kept as a file drawer full of individual cases with little organization or subsequent use. Individual airlines, sometimes through their labor union, also have access to the FAA Letters of Investigation (LOI) and violations of regulations by their maintenance technicians. Of course, local FAA Certificate Management Offices (CMOs) also have access to the violations or potential violations committed by the certificate holder, but these files have not often been examined for the purposes of risk management by the airlines. Even if the FAA files were to be examined, they would not provide any causal data on violations because the FAA inspectors report only the exact regulation that was violated and not the causal factors leading to such violations (Patankar, 2002).

A challenge in aviation maintenance today, is whether these databases and other information sources are complete and accessible. A large part of the answer is whether using information sources for risk management is part of the maintenance organization's design—whether they are built into the organization's structure, its process, its reward system, and its social roles. The sociotechnical systems approach is based on these principles of performance by design (Taylor & Felten, 1993; Drury, 1998), and which can be directly applied to risk management. Both the Concept Alignment Process (CAP) and the Aviation Safety Action Program (ASAP) are other examples of risk management 'by design'. Both CAP and ASAP are described further in Chapter 4. However, it is instructive here to highlight how organizational design can specifically include a database, and ASAP provides us with a good example. ASAP is a recent FAA-supported program, which is just emerging in maintenance organizations (FAA, 2002). ASAP requires a memorandum of understanding (MOU) between the maintenance management, the FAA, and the maintenance employees of an aviation maintenance organization. That MOU specifies a permanent tripartite *structure* (the Event Review Committee or ERC), it specifies a *process* of regular face-to-face meetings which uses consensus and discussion among participants, it specifies *roles and expectations* (maintenance employees who will voluntarily report their unintentional errors, and ERC members who will review the lessons from these reports in order to eliminate future errors) and it specifies *rewards* for new behaviors in the form of amnesty or reduced punishment for voluntary event reports that are included in the ASAP program. The 'how' of using ASAP information in risk management and process control is also contained in the MOU between the parties—it specifies that (a) the ASAP manager must maintain a secure database of de-identified event reports, (b) the FAA ERC representatives must conduct periodic audits of the ASAP database to confirm that any changes planned in response to individual reports are effectively implemented, and (c) the ERC, a whole, must affirm that trends noted in the error data are addressed through appropriate risk management techniques.

What do we Learn from Safety-related Information?

The bulk of what is learned about safety-related information is concerned with the social dynamics at the maintenance work place. We have already said previously that most airline maintenance facilities are not places where mechanics are encouraged to engage in open communication about safety or risk management. When there is a serious effort, such as ASAP, to listen to reports of safety hazards and violations, then 'what' is learned can be much richer in information and can help manage risk. Sociotechnical organizational design that leads to flattened hierarchies, use of teamwork, and open communication channels together with encouragement, and incentives for controlling risk in real time, can lead to higher trust among employees and management and thus to more information about risks and how to manage them.

The 'how' and 'what' are essentially the issues of awareness and understanding. The 'who' and 'why' of risk management begin to lead us toward action and implementation.

Who Should Learn about Safety-related Information and Why?

'Who' learns about safety related information and 'why' they should learn about it are both issues of social power, which cannot be ignored from a sociotechnical systems (STS) perspective (Badham, Clegg, & Wall, 1999; Coakes, Willis, & Clark, 2002). As individuals and professionals, maintenance technicians and their supervisors need to be able to take advantage of their own abilities to speak up and be heard on safety related matters. The 'who' should include all of them, and they should keep asking the 'why'. It is only when they do these things that they begin to put into action the 'how' and the 'what' they understand about their own maintenance organization.

In order to reduce risk, aviation maintenance strategy today should optimize the contributions of both people and information through technical support. This chapter discusses the social and technical issues related to understanding and harnessing that knowledge. It describes the progression from awareness of a problematic system to the implementation of a system designed for high performance. The STS approach is concerned with identifying and creating the conditions that lead to an optimum fit between people and their technology.

Sociotechnical Systems (STS) in Aviation Maintenance

STS analysis combines a technical systems view with a social systems view to capture the strength of both. The open system (one that is open to external influences) that results from this combination resides in a larger context or environment with which it must interact and to which it must adapt. STS is a theory and practice of organizational development in use for over 50 years and applied in a variety of industries worldwide. The aviation maintenance environment is a classic example of a sociotechnical system: the aircraft, hangar, and tools represent the technology while the people and processes used to facilitate maintenance actions represent the social system—together, they represent a sociotechnical system that is expected to seamlessly or harmoniously interact and accomplish the maintenance tasks.

The key features of the open sociotechnical systems framework are that work or production operations should be seen as follows:

- *A system* with interdependent parts.
- *An open system* adapting to and pursuing goals in external environments.
- An open *sociotechnical* system possessing an internal environment made up of separate but interdependent technical and social sub-systems.
- An open sociotechnical system with *equifinality* i.e., in which system goals can be achieved by different means. This recognizes the existence of organizational choice in the type of work organization that could complement any given technology.
- An open sociotechnical system in which performance depends on *jointly-optimizing* the technical and social sub-systems i.e., where neither the technical nor social sub-systems are optimized at the expense of the other (Badham et al., 1999).

Understanding any maintenance organization as an STS, and applying that understanding to improve risk management and safety performance requires three steps. These three steps or parts of analysis are the system scan, the technical analysis, and the social analysis.

The System Scan

The scan involves clearly defining the system's purpose, its value, its objectives, its boundaries and its salient environment. The scan helps members understand the system's organizational culture. In a maintenance organization its mission should be consistent with the overall mission, and culture of the larger company and that maintenance organization should examine the degree to which there is a common language and common product for shared purpose and culture. Examples of possible questions revealing airline mission (and to a lesser degree its culture) include the following: Is the airline the 'biggest', the 'cheapest', the 'safest', the 'most luxurious', or perhaps the 'most reliable'? Public utterances of such mission statements (including communication to employees) are extremely rare, so we will not often find them in company advertisements or in banners on the walls. Some airline missions are directly addressed in terms of maintenance performance objectives, but most are not. Of course safety and aircraft quality are maintenance deliverables, but it is infrequent that any air carrier has single-mindedly and exclusively pursued these ends alone. It is logical to expect that maintenance remains 'the place where highest mechanical safety is pursued at the most reasonable cost'. Whether a mission like that can and does become a unified and conscious element in a company's overall purpose is a matter that STS thinking and application can help with.

The Technical Analysis

The technical analysis focuses on the conversion process. It does this by defining and attending to the system 'throughput' (as in 'input-*throughput*-output'). It explores the question: 'what happens to our aircraft as they pass though our maintenance systems again and again?' The answers are important to understanding the maintenance function as part of a sociotechnical system because they focus on the *results* of the work performed rather than on the qualifications and qualities of the parts of the system expected to do the work. At the core of technical analysis and redesign lies a view of technology as more than instructions and equipment. The technical sub-system is seen as a conversion process transforming system inputs (the aircraft or its components) into outputs, under the control of skilled mechanics operating in social groups.

Simply conceived, the STS conversion process or throughput is quite different from the traditional 'work process analysis' of Industrial Engineering. Today's Industrial Engineering is a direct descendent of the early 20th century practice of creating man-made, ordered, rationalized (in mechanical or economic terms) structures and relationships. The time and motion studies of Frank Gilbreth (1911); F.W. Taylor's 'Scientific Management' (1911); and the 'Man-Machine Systems', and 'Human Engineering' of Alphonse Chapanis (1965) all supported the conclusions that efficient organizations assumed that workers were more like machines than beings of higher intelligence. Applied to aviation maintenance organizations, such rationalized organizations would consider mechanics as machines (sometimes complicated, unreliable, and recalcitrant machines) which are obliged to follow fixed instruction sets such as aircraft 'job cards' or 'maintenance manuals', or they are considered mere physical extensions of the technologies they use.

There is a real danger today of aviation maintenance inadvertently slipping backwards into those older, mechanistic views of Industrial Engineering if the 'process analysis' applied is derived from principles of Industrial Engineering or its most recent manifestation, 'Business Process Reengineering'. Business Process Reengineering, or 'reengineering', for short, seems like STS because of its application of a 'process analysis' to gain agreement and focus on the core business stream, but the similarities end there. Reengineering combines traditional Industrial Engineering methods and assumptions with the application of computer-based information technology to improve business functioning. Although the original reengineering text (Hammer & Champy, 1993) mentions human values and business purpose in passing, they are clearly neglected in application. Reengineering does not consider the social system of organization in any depth. It pays little heed to the social reality that most human work contains substantial interaction—and it ignores the corollary that role expectations provide more useful understanding of current work behaviors than do job descriptions of individuals. Reengineering is not a participatory approach to process analysis and is usually performed by external consultants with limited involvement by organizational members. The reengineering focus does not ordinarily derive from a systemic purpose and too often it becomes an examination (in the style of traditional Industrial Engineering) of the work process as a combination of human tasks and activities together with technology.

Although AMTs/AMEs have largely been spared the rigid structures and precise workflow of Industrial Engineering, the growth of 14CFR §145 aviation repair stations and their increased use of uncertificated mechanics offers modern efficiency experts a more willing audience for their ideas. If aviation mechanics in repair stations are currently under the efficiency microscope, the examination of certificated mechanics in airlines is a likely next step. The central question for risk managers to ask is as follows: Does the process analysis being used, treat mechanics as 'task followers' or does it expect mechanics to be 'variance controllers'? Task followers would be machine operators or humans in a technical system with minimal opportunity for the use of their intelligence. Variance controllers, on the other hand, would be intelligent humans who are tasked with the freedom to make appropriate changes to the work process in order to maintain the functional priority of the system—in the case of aviation maintenance, delivery of a safe aircraft.

Do not be tempted to import the Industrial Engineering assumptions in aviation. If airline maintenance work process is subjected to process analysis that pays more attention to individual task activities than to controlling important conditions in the throughput, beware. An STS technical process analysis, on the other hand, focuses on key variations in the throughput—the aircraft being transformed. Furthermore, it requires human interactions—that is, people with their ability to understand and control key variances when they occur in the process. Moreover, sociotechnical systems rely on role definitions where mechanics *expect* dialogue among themselves and their management in order to control those variances most effectively. Such a dialog not only makes the human fallibility acceptable, but also provides mechanisms to maintain the overall reliability of the system. The human and technical reliability perspectives are discussed in more detail in our previous volume (Patankar & Taylor, 2004, Chapter 7).

In sociotechnical analysis, the technical system's conversion process requires the *intelligent and adaptive interaction* between humans and tools, and between humans and humans, to follow work rules as closely as possible given ever-present variations in the material, the tools and the circumstances within the context of organizational mission. For over a century our industrial-factory society has allowed technology to define and drive people, as opposed to creating workplaces in which people drive technology in pursuit of the organizational or professional mission. To meet their mission, people require an overview or big picture. The systems scan, and technical system analysis, within an STS context provides that overview and helps empower mechanics to manage their work for outcomes, instead of just doing the tasks to which they were assigned. When those outcomes are related to safety of flight, maintenance professionals can be more responsible and more responsive to what needs to get done. This allows the needs of the product to drive the tasks and activities as opposed to narrowly defined task requirements driving the work process.

Before we proceed further, we would like to caution the readers that we are not advocating procedural violations if the mechanics believe that there is a better way to accomplish the task. In our opinion, the procedure-driven attitude in aviation can be traced back to the World War II era. At that time, large volumes of maintenance personnel had to be trained in a very short time. In order to achieve that task, most mechanics were taught just the core skills and aircraft-specific knowledge was reserved for the respective maintenance manuals. Subsequently, the FAA inducted these maintenance instructions into the regulatory framework in order to facilitate the regulatory oversight of civil aviation maintenance operations. Consequently, we now have maintenance instructions for most aircraft-specific tasks, and all such instructions must be complied with in accordance with 14CFR §43.13. Such a tight coupling between maintenance instructions and regulatory compliance tends to predispose mechanics into regulatory violations if they were to deviate from published maintenance instructions, even when such instructions are wrong. The ASAP, described earlier in this book, is beginning to address this challenge. As a result of this unique process, it is now possible to control variances without the fear of disciplinary action from the employer or enforcement action from the FAA (c.f. Patankar & Driscoll, 2004).

In aviation maintenance, the analysis of an STS conversion process can be seen to proceed through a number of stages or process steps, each of which represents a change in state of the aircraft from intake to return-to-service. The aircraft, during these 'state changes' are subject to variations or variances in their condition. These variances are sometimes expected, but they are often unpredictable as to time and extent. The most important variances, the so-called 'key variances', have to be controlled (or controlled for) if system goals are to be effectively achieved (Drury, 1998).

Examples of technical system variances found in any commercial aviation maintenance system include the following:

- Nature and/or extent of flaw, defect, or damage tends to vary. More complexity or extent of damage may require special skills, or more coordination with other maintenance employees and it will surely require more time.

- Large and complex repairs such as cracked door frames or extensive corrosion of floor structure and pressure bulkheads often define the 'critical path' of heavy maintenance work and become key variances in that setting (Taylor, 1991).

- Visibility of flaw, defects, or damage tends to vary. Sometimes, without proper or complete cleaning, flaws are less detectable and as such can be key variances (Drury, 1998). Less visible flaws are often detected late, if at all, and can delay, prolong, or defer other priority work. Ironically, knowing that flaws such as corrosion are often hidden does not often translate into preventative maintenance programs such as inspecting and treating corrosion-prone areas, to control these variances in advance. Instead extensive corrosion is sometimes not detected until late in a maintenance project which exacerbates the effects of optimistic time estimates and poor coordination/communication between departments and shifts (Taylor, 1991).
- The time required to perform similar repairs may differ. Specific defects and damage require different time to return the item to service. The longer the time required, the greater priority to begin work.
- Key variances are not always the most time consuming repairs, but they may demand exotic parts, or special engineering planning, or intricate scheduling of other repairs—and these things can add to the time required (Taylor, 1991).
- Time allowed for repairs may vary. In flight-line maintenance in particular, the time allowed is often less than the time required, and complete repairs must be deferred until a later time.
- Parts availability may vary. Identifying what needs replacing early enough can permit cost saving or preclude delay of repair beyond the normal estimate.

Once specific variances are identified as critical to attainment of the maintenance system's mission, they can be examined more carefully to determine (a) *why* they are declared important and (b) *who* uses *what* type of information to control them and *how*. Variances so identified become 'key variances' for which system design can be addressed (Drury, 1998). At this point, the degree of risk to flight safety of each key variance can and should also be assessed. With the increasing use of maintenance error databases from ASRS, MEDA, and ASAP such risk assessment can be undertaken using probability estimates and statistical modeling (Patankar & Taylor, 2002; Marx & Slonim, 2003; Kanki, Marx, & Hale, 2004)

The Social Analysis

In aviation maintenance, the social system analysis examines the quality of work-related communication among mechanics and others. It is an evaluation of who talks to whom, about what, and how it is working for the overall system. Social systems analysis is linked to the technical analysis because the most important communications are about the technical 'throughput' (i.e., the aircraft—including its defects and scheduling requirements). But the social system is also the wider mechanism for empowerment and employee participation in risk management.

The social system in a maintenance organization includes the mechanics (AMTs/AMEs), members of their work group, members of other work groups, their supervisors, other shifts, other departments, higher managers, government inspectors, as well as their contact with flight operations, customer services, and mechanics in other companies. The social system is described as serving four functions:

- *Goal attainment* is the control of key technical variances.
- *Adapting* is the flexible response to a changing environment [what has been called 'requisite variety' (Ashby, 1956)—recall the Hawkins-Ashby model we discussed in our earlier volume (Patankar & Taylor, 2004)].
- *Integrating* refers to those relationships promoting cooperation and managing conflict.
- *Long-term development* provides those human interactions that are necessary for sustaining the system over time.

An easy way to remember these four functions is that their acronym is GAIL. GAIL or the four functions, also helps us see, respectively, the 'how,' 'why,' 'who,' and 'when' of sociotechnical system behavior.

The social analysis focuses specifically on the primary relationships of a 'focal role'. They are the 'focus' because it is they who are directly responsible for the timely delivery of a high quality product (Taylor & Felten, 1993, pp. 112-114). In aviation maintenance, that role is entrusted to the AMT/AME—the licensed maintenance professionals who actually do the work and/or certify that the results are airworthy. These are the individuals in the system who are most responsible for risk management. There is no question that maintenance managers are also responsible for risk management, but the AMTs/AMEs doing the work and certifying subsequent airworthiness are the most central to safety, and they should be equal partners in being responsible for safety. In the sociotechnical analysis, their communication with others is viewed through the lens of the GAIL model.

The following sociotechnical view of aviation maintenance social system and culture was studied in 1990 (Taylor, 1991). Since that time, we have had ample opportunity to confirm those early observations in the United States as well as in other countries (c.f. Taylor & Patankar, 1999). That early, documented, social system analysis of heavy maintenance included observations and interviews with some 250 AMTs, maintenance supervisors, and managers (plus over 80 degreed engineers and other aviation support professionals) in eight US air carriers and repair stations. That analysis showed that among all four social functions, goal attainment ('G' in the GAIL model) accounted for most AMT communication, but it also revealed that in 1990 only about 20% of those contacts were positive or contributed to satisfactory control of key technical variances. It was discouraging to note that 40% of the total exchanges were negative and tended to disrupt effective variance control. The remaining 40% of the goal attainment communication was neutral—neither enhancing nor diminishing the control of critical variances in the maintenance process.

Likewise, many interactions were observed that contributed to the integration function ('I' in the GAIL model), but unfortunately over 60% of these represented negative incidents—that is, they were disruptive or disintegrative to the social fabric of the maintenance organization.

That study did not find much of the mechanics' social interaction contributing to either adaptation to a dynamic external environment ('A' in the GAIL model) or to long-term development ('L' in the GAIL model). Together, communication coded in these two categories amounted to less than 20% of the total. What was observed as 'A' or 'L', however, was more neutral than negative in its effect on system effectiveness.

These results paint a picture in which mechanics and inspectors play a central role in accomplishing the essential task of variance control ('G'), with considerable guidance from their foremen, with some cooperation from others in their work group, and much direct contact (although a high proportion of it negative in outcome) with other employees in the maintenance system. Further, it shows that mechanics' cooperative work relations ('I') with co-workers in the same occupation are not well developed. Mechanics have an individualistic style. We described in our earlier volume that despite national differences within occupations, maintenance people (as a total group) are more individualistic than any other occupation in commercial aviation (Patankar & Taylor, 2004 Chapter 1). Mechanics are proud of their macho style and their individual certification, and this personal responsibility has an added effect of making mechanics and their supervisors less willing to share work across shifts, or with less experienced or less skilled colleagues. The resulting performance can be slower and the ability of mechanics to exchange ideas or information is limited. This restricted communication further supports traditional emphasis on the individual contributor as the basic work unit. From a sociotechnical systems view of risk management, this situation leads to a lower sense of collaboration, lower mutual trust, and a diminished safety culture (a more extensive treatment of this is contained in Chapter 5 in the present volume).

At those sites where (by labor contract) lead mechanics were in strict charge of work assignment and operational guidance, mechanics' contact with foremen is much reduced and results in generally less effective coordination between shifts because foremen, no longer in direct contact with mechanics, are still responsible for formal shift turnovers. The quality of technical guidance ('G') can suffer as well because high seniority mechanics sometimes bid into lead jobs without sufficient breadth of technical experience to always understand the work they are assigning to others, and the results of which they are describing to foremen. It is obvious that mechanics in lead positions are disadvantaged by a lack of formal leadership training—something many, if not most, foremen are exposed to. Lead mechanic's acute awareness of this lack of 'people skills' ('I') has been quantitatively documented (Predmore & Werner, 1997; Taylor & Christensen, 1998, p. 14). In a sociotechnical systems view of risk management, this situation leads to low levels of positive and open communication, lower mutual trust, resulting in a less positive safety culture.

Little communication in the service of adapting to outside changes ('A') is a common condition in aviation maintenance. Normally it is represented by mechanics' ignorance of, or apparent disinterest in, the company's 'big picture'—sometimes with the view that with the pressure for faster and faster work, 'the best we can do is "good enough not to have an accident"' (i.e., near misses do not count). At its worst, mechanics' view of the outside world is of the company's decline in a shrinking employment market—and in that case, their communication takes the form of mutual commiseration as well as speculation about job opportunities elsewhere.

A small role by management in training, or personal long-term development ('L') was noted in 1990 and subsequently. Most mechanics say they obtained on-the-job-training (OJT) from more senior employees;

however, there are numerous situations where a lack of qualified senior employees diminishes the quality of this training. At one of our research sites, we observed that most of the experienced mechanics were on the day shift and most of the junior mechanics were on the night shift. Consequently, OJT or knowledge transfer from experienced to novice was almost impractical.

Cooperation (and consequent respect) among maintenance employees between departments are especially poor when the organizational structure is arranged so that maintenance operations, materials, component shops, inspection, training, quality assurance, and planning/scheduling functions all report to separate vice presidents. When these departments have competing goals and diverging or incomplete understanding of the maintenance and company mission, problems can escalate near the limits of the system's ability to cope. This structure creates boundaries through which it is difficult to communicate, leading to more finger-pointing and promoting more politics than productivity. From a sociotechnical systems view of risk management this situation leads to a lower sense of shared purpose, lower employee morale, and a diminished safety culture.

In the report of the original study the following four recommendations were offered for improving aviation maintenance as a sociotechnical system (Taylor, 1991):

- Increase the workforce competence.
- Emphasize and support maintenance system centrality in company purpose.
- Develop commitment to human values that reflect the desired practices of management and employees, and which enhances the logic of those practices.
    - Consistency in values and practice and an open attitude to communication with employees creates greater commitment to the company and mutual trust among its members.
- Create and endorse teamwork in the maintenance system.

The above recommendations are as timely now, over a decade later, as they were at the time. The main additions we would make would be as follows:

- Apply more careful methods to assess the probability of safety of flight degradation to any key variance identified in the technical process analysis.
- Specify behavioral outcomes in addition to attitudinal ones.

With regard to the first of these, both Patankar & Taylor (2002) and Marx & Slonim (2003) have offered methods to add probability estimates to event data, even when actual rate data are not available. The second recommendation has been addressed in the approach to self-reports in questionnaire and interview surveys we have validated and reported (Taylor & Patankar, 2001).

Implementation

Sociotechnical systems have been successfully and widely implemented in many organizations for over 50 years (Taylor & Felten, 1993; Coakes, Willis, & Clarke, 2002). Although there is an interest in applying sociotechnical practice to aviation maintenance (Kanki, Max, & Hale, 2004) there are few documented examples. Several hazards previously described in detaïl (Taylor & Robertson, 1995; Taylor & Patankar, 2001) await managers and executives contemplating any serious human or social system changes. Chief among these are a lack of (a) serious and long lasting organizational commitment to the change and (b) involvement of people who do the work being changed in the analysis and design of the changes. It is important to distinguish between *organizational* commitment mentioned above in item (a) and *management* commitment. In our past research (Patankar, & Taylor, 2000a), we have observed that managers tend to change assignments every 3-5 years. So, if only certain managers support the change program, the change is not likely to maintain its momentum after its champion has left.

During the periods of MRM implementation it has been observed that when maintenance personnel with a high level of experience from one airline have relocated to another airline, they tend to locate fellow expatriates in the new organization and compare the past with the present, often at the expense of their present employer (Patankar, 1999). Patankar called this pattern of past experience 'sub-organizational mosaic', or SOM. This SOM construct was subsequently empirically tested and confirmed (Patankar & Taylor, 2000)

Changed attitudes include initially high enthusiasm for MRM programs. For one airline studied, over 95% liked the MRM program immediately after training and thought it would be useful. However, six months later, this endorsement dropped to between 75-80%. Six months after training, trust in one's supervisor's safety practices also decreased markedly from post-training levels. The value of trusting coworkers also reverted to pre-training levels (Taylor & Thomas, 2003b). It was noted that the company initially trained those AMTs with extensive experience in other airlines. These initial trainees fit the conditions for SOM. As the training's vanguards, these high-SOM individuals were responsible for communicating their initial impressions of the MRM program to their coworkers. Equally important, this same group was asked six months later (not incidentally during a time of heated contract negotiations) to verbalize their subsequent impression of the program. Their frustration and discouragement with the pace of the program, combined with the invidious comparison of the company with their former employers, acted to produce some very negative views.

Implementing Effective Change

If the 1990's Maintenance Resource Management (MRM) was for *individual safety awareness*, the new millennium's STS-MRM is for *systemic and proactive risk* management. In this decade, management concerns have expanded to include not only the hangar and line maintenance, but also the flight deck, passenger cabin, the ramp, and the ticket counter. Sociotechnical systems ideas and methodology will be suffused into the MRM practice ever more in the 21st Century.

The Three Pillars of Change

Successful organizational change can be defined as sustained positive effects on intended end results that are diffused throughout an enterprise. Wherever it is applied, MRM is becoming a fundamental source of organizational change and improvement in aviation. To be successful, organizational change through STS requires three elements to be present: (a) unequivocal top management support and vision of the purpose for the change, (b) a well-conceived and relevant intervention, and (c) timely appropriate feedback, through a broad range of measurement and evaluation techniques. The necessity of these three aspects in successful STS in other industries is well known and has been described elsewhere (Taylor & Felten, 1993).

The Challenge in Maintenance

Today's aircraft mechanics, inspectors, and foremen face an extremely high-tech maintenance task. These individuals have successfully learned to work with fly-by-wire, fly-by-light, the 'glass cockpit', Built-in Test Equipment (BITE) and other computer-based tools, as well as with composite structures, new repair procedures, and ever increasing legal obligations. Maintenance *people*, however, are delayed in their rapid deployment of this technical learning because they have not mastered how to successfully communicate. Unlike cabin personnel and gate agents, who self-select and are selected because of their skills and ability to interact with others, mechanics select their occupation because they want to work with tools and with large and complex machinery. But, the complexity of today's civil aviation industry demands that mechanics (as well as all other aviation occupations) learn to communicate effectively among themselves as well as with other groups. Each of the three elements required for effective organizational change apply to Maintenance Resource Management programs.

The Three Pillars of Change for STS in Maintenance

1. Management support: Successful change requires unequivocal top management support, or making STS part of the culture. What happens when a dedicated, inspired Senior Technical Operations Vice President creates an MRM program for all 2,200 of his management and staff support personnel? A well-known case has shown that if that executive dedicates himself to that vision long enough, if he is persistent in his visible sponsorship, and if he is clear in his conviction that scientific evaluation of the program will improve its acceptance and continued development as well as validate his vision, then results actually happen (Taylor & Robertson, 1995). His maintenance managers begin to seriously value the open, assertive communication, safe work habits, and problem-solving methods that the program espouses. Those managers go on to believe (and report in surveys and interviews) that this program, unlike most others they have experienced, will stay long enough 'to make a difference'.

2. Quality intervention: Successful change requires a well-conceived and relevant intervention. Proactive risk management programs thus created have been observed (for an illustration of one of these, see Patankar & Taylor, 1999b) and are well planned and efficiently executed. They are marked by appropriate new structure and processes designed both to initiate proactive communication and to support and reward its continuation as a prime behavior involving mechanics in managing risk.

3. Measurement and feedback: Successful change requires timely and appropriate feedback—through a broad range of measurement and evaluation. Interpersonal trust, for example, is so important to any such program that it is *essential* to measure it as an enabling condition before change is undertaken because without trust there is little willingness, or likelihood, to try something new and uncertain. Measuring trust again, once intervention of a high-quality MRM program has begun, is also useful to understand how this essential condition has improved. Measuring behaviors, which mechanics and managers expect to change, are essential to confirm that the intervention program is succeeding and to provide incentive for further change. Measuring probabilities and risks of errors have already been mentioned earlier, and they are also essential—both as benchmarks to identify the safety needs of an enterprise and to track improvements in the goal attainment produced by the change.

Maintaining and Diffusing Change—Dancing with the Bear

Can systemic change programs like STS be assured of continued success? The answer lies in all three fundamental elements of change. If it does not evolve, it is likely to become a one-shot 'project', the positive effects of which will be arrested. If the program loses its management support, a gradual reversal in effects is likely. If that support is lost suddenly, the reversal is accelerated and measurable decay is virtually guaranteed.

The cessation of evaluation is the least important of the three pillars (at least in the short term), and the absence of negative results during reversal may even prolong the illusion of continued progress. Without feedback, however, even the most effective program will gradually drift in the direction of presumed strengths which in reality can become the source of complacency and arrested development, if not decay and reversal. The effects of frequent measurement on program visibility and evidence of its continuance is an immeasurable benefit at relatively small cost. The opportunity measurement provides for participant dialogue with facilitators and top management also cannot easily be replaced.

In addition to the three pillars of change, we cannot escape the effects of the *context* of change—the external environment. For example, changes in the market, customers, or competition can cause management to react with policies that seem to work at cross-purposes with the involvement and open communication and STS. Such ambiguity or conflicting policies require even greater clarity in already familiar programs such as MRM. These programs deserve that clarity of focus, especially during environmental turmoil, if they are to survive and succeed.

What permanent effects do an organization's changes have on its members? Nothing is totally neutral or completely reversible. An effective program cannot survive without support, content, and feedback, but its effects are not fragile either. Unfortunately, these effects are often a 'boomerang' that comes back to the enterprise in the form of negative attitudes. The abandonment of such a program results in discouragement, despair, and cynicism. The loss of employee motivation because of the boomerang effect can have a considerable effect on performance. As the story goes, one cannot stop dancing with a bear merely because one wants to, but only when the bear stops.

Recommendations

Elsewhere, we have offered evidence that many MRM programs did not reach their full potential due to lack of management follow-up. Studies have shown that many aviation maintenance managers experience high mobility, and therefore they may not have the extended time required to consistently support the MRM programs they find themselves involved in (Patankar & Taylor, 2000). It has become clear that MRM programs should be integrated with the core organizational purpose if they are expected to survive more than a few years.

We have been vigorous and visible in offering an approach to make MRM programs more independent of the changes in management and the effects of mechanics' past organizational experiences. That approach is to have a documented and integrated human resources master plan which includes maintenance and which is approved by the President/CEO of the airline (Patankar & Taylor, 2000). This plan should clearly identify the anticipated outcomes of the program, associated time-line, and budget. It should be results-driven. All management personnel should then be held accountable for abiding by the plan. Unless all management personnel are evaluated for their implementation of the master plan, the human factors program may not get consistent support during management succession, or if and when present management suddenly embraces a new program. We advocate that in advance of such events, management should mandate the behaviors and goal attainment of MRM programs. Employees appreciate the MRM programs once they have been involved and their initial enthusiasm and favorable attitudes are literally universal. They want to believe in the MRM message, but need guidance and leadership to behave effectively. Data show that management appointments change every three to five years, so it would not be prudent to think that a certain favorable manager would be able to support an MRM program forever. Program champions are essential to initiate the training and implementation, but these champions must also make sincere attempts to make the program independent of themselves.

We have spoken about 'three pillars' of successful cultural change in aviation maintenance. Those three pillars are (a) unequivocal top management support and vision of the purpose for the change, (b) a well-conceived and relevant intervention, for behavioral change, and (c) timely, appropriate feedback through a broad range of measurement and evaluation activities.

The evidence presented in this chapter continues to add to the truth of these three pillars. They continue to be valid. The contribution presented here is that relevant and effective behavioral intervention is required in addition to appropriate, self-administered measurement tools for appropriate feedback.

Chapter Summary

Using sociotechnical systems (STS) provides a methodology for understanding and improving risk management. STS helps us understand our work process and helps us improve our practice of proactive safety. In part, STS does this by helping us see what downstream hazards can result from our actions in the overall work process, and likewise, how others 'upstream' from us in time or technical action are unwittingly creating hazards or traps for us. STS also provides guidelines for participants (e.g., system members) themselves to change the structure (e.g., groups), the infrastructure (e.g., tools or technology), and the culture (e.g., individual and group incentives, communication skills).

Review Questions

1. Describe why awareness-type maintenance human factors programs tend to be viewed as 'flavor of the month'?
2. What strategies do you suggest to fully integrate maintenance human factors into the organizational safety culture?
3. What is a sociotechnical system? Give examples from fields other than aviation.
4. Why is it important to optimize the technical as well as social aspects of a system? Give examples of successful as well as unsuccessful optimizations—use personal experiences and outside reading to illustrate your points.
5. Describe some of the key challenges in minimizing maintenance errors.

Chapter 3

Pre-Task Analysis

Instructional Objectives

Upon completing this chapter, you should be able to accomplish the following:

1. Recognize that there are a multitude of pre-task symptoms that need to be recognized and managed effectively in order to minimize the probability of errors.
2. List some examples of how lack of knowledge, lack of tools/equipment, time pressure, and ineffective shift-turnover can contribute toward errors.
3. Describe your role as a mechanic/manager in utilizing pre-task analysis to minimize maintenance errors.

Introduction

Pre-task analysis is essentially a means to take a moment and assess the task at hand: Are you physically, emotionally, and intellectually prepared to perform this task? Do you have adequate resources? How do you plan on minimizing the probability of committing an error *today* and *at this task*?

Even if you look at a handful of randomly selected ASRS maintenance reports, it is clear that maintenance professionals work under physically strenuous conditions, they tend to be distracted or preoccupied by emotional issues (like most normal humans) such as family problems or job uncertainty, and/or are sometimes asked to perform maintenance actions for which they are not trained or for which they do not have approved tools. How many of the above conditions are true will depend on the economic health of the maintenance organization and the extant safety culture; nonetheless, the challenge to exhibit professionalism and pride in spite of varying levels of de-motivating factors is a reality.

In this chapter, we build upon the pre- and post-task analyses that were introduced in our previous volume (Patankar & Taylor, 2004). Here, we present *preparation, work management,* and *communication* as the three fundamental elements of pre-task analysis. Each section presents some specific steps that you can take to help yourselves accomplish your maintenance tasks with a lower probability of error. If some of these steps seem like 'common sense' to you, please allow us to reassure your common sense.

Preparation

The basics of preparation include self-assessment of intellectual and physical capabilities and availability of resources. For example, consider that you were tasked with accomplishing the repair of a composite flight control surface. Your immediate response should involve an assessment of whether or not you have the intellectual, physical, and resource capabilities to accomplish the task. Are you trained to do such work? In the past, have you done this type of repair to your supervisor's satisfaction? Are you feeling physically fit to perform this repair? Do you have the resources to perform this repair? If answers to all these questions are satisfactory, you may proceed to think about all the things that could go wrong and cause you to commit an error. For example, are you using the correct maintenance manual reference? Do you have the appropriate temperature and humidity conditions to accomplish this repair, especially if it is to be done outside the hangar?

Remember the Hawkins-Ashby model we discussed in our previous volume (Patankar & Taylor, 2004)? Well, here is the synopsis of that model. Consider the SHEL components (**S**oftware, **H**ardware, **E**nvironment, and **L**iveware) (Hawkins, 1987). Are any of these components different from what is specified in the maintenance manual? For example, is the repair to be performed indoors and you are forced to do it outdoors? In this case, the *environment* has changed; therefore, according to the Hawkins-Ashby model, one or more of the remaining components have to change so as to accommodate this variation in the environment.

From the perspective of pre-task preparation, the key areas of emphasis are as follows: knowledge/skill level, adequacy of resources such as maintenance instructions, tools, parts, equipment, etc., number of distractions, level of time pressure, and physical fatigue.

Do You Have Adequate Knowledge to Perform the Assigned Task?

Although Title 14 of the Code of Federal Regulations (14CFR) requires that each mechanic should have performed a particular task under the supervision of, and to the satisfaction of, an appropriately authorized mechanic or FAA Administrator/Inspector, we continue to find mechanics forced into situations where they are given a task for which they have received no training. Fundamentally, the organization is in violation of either 14CFR §121. 375 or 14CFR §145.55, depending on whether it is an airline or a repair station, and the individual performing the maintenance (if the individual is a certificated mechanic) is in violation of 14CFR §65.81(a):

> 14 CFR §121.375 Maintenance and preventive maintenance training program.
>
> Each certificate holder or person performing maintenance or preventive maintenance functions for it *shall have a training program* [emphasis added] to ensure that each person (including inspection personnel) [emphasis added] who determines the adequacy of work done is *fully informed about procedures and techniques* [emphasis added] and new equipment in use and *is competent to perform his duties* [emphasis added]
>
> 14 CFR §145.55 Maintenance of personnel, facilities, equipment, and materials.
>
> Each certificated domestic repair station shall provide personnel, facilities equipment, and materials at least equal in quality and quantity to the standards currently required for the issue of the certificate and rating that it holds
>
> 14 CFR §65.81 General privileges and limitations.
>
> (a) A certificated mechanic may perform or supervise the maintenance, preventive maintenance or alteration of an aircraft or appliance, or a part thereof, for which he is rated (but excluding major repairs to, and major alterations of, propellers, and any repair to, or alteration of, instruments), and may perform additional duties in accordance with §§65.85, 65.87, and 65.95. However, he may not supervise the maintenance, preventive maintenance, or alteration of, or approve and return to service, any aircraft or appliance, or part thereof, for which he is rated *unless he has satisfactorily performed the work concerned at an earlier date* [emphasis added]. If he has not so performed that work at an earlier date, he may show his ability to do it by performing it to the satisfaction of the

> Administrator or under the direct supervision of a certificated and appropriately rated mechanic, or a certificated repairman, who has had previous experience in the specific operation concerned.

Based on the clarity of these regulations, it is reasonable to assume that all certificated mechanics and managers of both airlines and repair stations are fully aware of basic training requirements. Then why do organizations place mechanics in situations that set them up for regulatory violations, and in so doing, violate regulations themselves? There are several interconnected factors that lead to such violations. From an organizational perspective, it could be a simple issue of economics wherein the organization simply cannot afford the training and is taking significant business as well as safety risk by compelling untrained or marginally trained individuals to perform maintenance. On a more complex level, it could be a case of organizational culture—pathological organization—wherein regulatory violation is a norm rather than exception (c.f. Westrum & Adamski, 1999). On an individual level, it may be a matter of survival. When there are no other jobs in the aviation industry, the mechanics are left with fewer options; few can afford to just quit their job because they do not agree with their organization's safety values. Under such circumstances, a handful of mechanics, typically senior mechanics, have challenged the company management and risked their jobs in an effort to improve their company's safety practices.

Example 3.1 illustrates an extreme example of lack of training and absolute disregard to safety at the organizational level. This is certainly an extremely rare situation; however, as you will note from Example 3.2, placing untrained or inexperienced individuals in specific maintenance responsibilities, by a supervisor, is not that rare.

Example 3.1: An extreme example of lack of training at the organizational level

ASRS Report Number 395238

(Public document, edited for clarity)

> During an FAA inspection, inspectors found that repair procedures may have deviated from the correct procedures. During the time that I have spent in the composite shop, I may not have complied with repair procedures completely due to lack of composite training. The company was made aware that it had not provided training in composite repair to specific aircraft type. Callback conversation with the reporter revealed the following information: Prior to the FAA shop inspection, very few people were trained on the repair and buildup of composite structural components. The reporter said that some of the work done by untrained

mechanics was not per the manufacturer's structural repair manual and was never inspected. The reporter stated that after the FAA shop inspection the company started a training program and now maintains training records of mechanic qualification. Callback conversation with ASRS Reporter Number 394876 revealed the following information: The reporter said the shop operated with production as the primary goal with repairs are not being done per the structural repair manual which cut job time and increased production. The reporter stated that there never was a training department and no training records were maintained. The reporter said a lot of unapproved repairs went out of this shop and were never inspected.

Example 3.2: An example of supervisors placing untrained individuals to perform maintenance

ASRS Report Number 409606

(Public document, edited for clarity)

[I was] installing a main left inboard tire assembly on air carrier ship xyz, a B737-200. The axle nut retainer ring was improperly installed. The maintenance manual states that the lock holes should align (between the nut and the spacer) after torquing. This was not done. The retainer was installed through the spacer hole only, not the nut. This condition could allow the nut to loosen and back off until a nut corner contacts the retainer. The manual is clear on this step, right after torque procedure and [I] simply didn't read it carefully enough. Also, I had never done the operation before (lack of training) [It is illegal to perform maintenance without supervision if one has not demonstrated proficiency in the past]. The aircraft was dispatched and no problems resulted before the condition was reworked. This was an honest oversight and I notified my supervisor as soon as I realized the installation was incorrect.

Do You have the Complete, Approved, and Up-to-date Technical Data?

Mechanics are always required to use the most current and approved maintenance instructions and they must trust that the computerized or paper-based maintenance instruction system is providing them with the latest and approved maintenance instructions. However, it is very common to find that mechanics consider maintenance instructions to be confusing, incomplete, or flat out wrong. Sometimes, there is a discrepancy between the instructions in a task card and those in the maintenance manual. Sometimes, the task is so routine that mechanics get too comfortable with the task and tend to do it without referring to the maintenance manual, setting themselves up to miss critical updates in the maintenance instructions.

As far as maintenance instructions or technical data are concerned, here are some norm-based errors:

1. *Use of IPC rather than Maintenance Manual*: Sometimes, mechanics assemble systems/subsystems by looking at a diagram in the Illustrated Parts Catalog (IPC), not the maintenance manual simply because the IPC has better diagrams and the mechanics believe that 'a picture is worth a thousand words'—it is much easier to assemble something by looking at a diagram than by reading the text instructions. While that may be true, IPC cannot be quoted as an approved source in the maintenance logbook and it is likely that one could miss certain critical caution and warning messages that are posted in the maintenance manual but not in the IPC.

2. *Assuming that a previously installed part or previously performed repair is correct*: In the industry, we have heard phrases such as 'what comes off the aircraft, must go back on the aircraft' and 'Oklahoma blueprint' [our apologies to the people of Oklahoma]. In the former case, the norm is that whatever came off the aircraft in a disassembly process, must be installed back in the assembly process. While this is generally a valid statement, it overlooks the importance of checking the validity of the previous part. What if someone else had installed a wrong part? The norm would encourage the new mechanic to reinstall the wrong part. Every disassembly and assembly should be regarded as an opportunity to validate part numbers and repair schemes. Of course, this is a time-consuming process and needs support from management. The latter issue is that of 'Oklahoma blueprint'. This means that one needs to make sure that the left side of the airplane matches the right. If one tire is replaced, make sure that it matches the other; if a repair is performed, make sure that it matches similar repairs on the same aircraft or other similar aircraft; and if a part is installed, make sure that it matches the part number of the one that came off. Well, again, these norms set the mechanics up for errors.

3. *'Return aircraft to normal'—The catch-all trap in aviation maintenance:* Just what exactly is 'return aircraft to normal'? At the simplest level, it could mean that you must close all panels and reset all switches, circuit breakers, etc. In reality, nobody knows what 'normal' means. Yet mechanics routinely sign-off this last item on their task

cards. In case it is later discovered that one of the crucial steps such as removing a gear locking pin or resetting a circuit breaker was missed, the mechanic could face regulatory violation charges.

Example 3.3 illustrates a classic case of confusing technical information. In this case, mechanics found a way to accomplish the repair required by an airworthiness directive using a 'better' repair scheme. The first mechanic who performed this repair, misinterpreted the AD and used a repair scheme that was different from the AD. Other mechanics assumed that the previous repair was not only good, but also that it must have been better than that specified in the AD. If they had not tried to seek some resolution on this matter, the unapproved repair would never have been detected.

Example 3.3: A classic case of confusing technical information

ASRS Report Number 343710

(Public document, edited for clarity)

> While working airworthiness [directive] xx-vv-zz on aircraft abc, I misinterpreted a portion of the airworthiness directive which addressed the wire routing and splicing of the fire bottle circuit. After observing several aircraft being done in this manner, aircraft zzz wiring was [also] routed and wired not to the airworthiness directive specifications. We thought the reason for the change being the elimination of cutting and splicing some wires. Safety was not compromised. All subsequent checks and tests were good. After three aircraft were wired in this configuration it was determined the letter of the airworthiness directive was being violated. I contacted my maintenance foreman and advised him of the situation. The foreman was asked to request engineering to revise the portion of the airworthiness directive dealing with the wiring. This would eliminate wire splices and ground wire replacement. He agreed to do so and contacted engineering at xyz and presented the problem. Upon returning he informed us that the engineer thought what we were doing was a great idea and saw nothing wrong with what we were doing. We were not advised if the wiring portion of the airworthiness directive would be revised. The foreman's instructions were to do things as we have been doing, that is not wiring the aircraft to the airworthiness directive specification. Another two aircraft were modified in this manner. Upon discussion with other shift mechanics we realized that while the intent of the airworthiness directive had not been violated, the letter and specifications were in violation. The lead mechanic contacted the engineer previously contacted by the foreman and was advised under no circumstances could we deviate from the airworthiness directive specifications. The engineer also related that the foreman was advised of this directive. When confronted with this fact the foreman thought it was

no big deal but agreed to help rectify the situation. He was given the tail numbers of the airplanes involved so a plan could be devised for correcting the wiring. Before the airworthiness directive note expired we talked with a director of maintenance and a member of engineering in xyz. It was agreed to bring the affected airplanes in that weekend for correction before the airworthiness directive expired. We were advised that working the weekend may be required. The weekend passed and we were again advised that things had been taken care of or resolved. This was Jan/mon/96. On Jul/mon/96, six months later, we were sent out to perform this airworthiness directive compliance on aircraft zzz. This should have been done in Jan/96.

Callback conversation with reporter revealed the following information: reporter was assigned to work the wiring portion of airworthiness directive xx-vv-zz on aircraft abc, a B-737-400, fire extinguisher circuit rewiring. Reporter misinterpreted the job card instructions on wiring splices and routing of the wire bundles. The completed work was not as specified in the airworthiness directive job card. It was the mechanic's opinion the wiring job they had done was a better installation and avoided cutting wire bundles and adding splices. The mechanics doing the work requested the maintenance foreman to contact the engineer who wrote the job card to allow some deviation from the airworthiness directive specifications. The foreman advised the reporter and mechanics involved that the engineer had been contacted and it was approved to do the job their way. This later proved wrong. The engineer would allow no deviation from the job card specifications. This airplane was eventually rewired six months after the airworthiness directive expired. No explanation was given to the reporter on why the delay to rewire this airplane and six others wired incorrectly. Reporter has 22 yrs experience as an A&P mechanic. Reporter was not contacted by the FAA.

Do You have the Appropriate Tools/Equipment to Perform the Assigned Task?

Aircraft mechanics are often considered 'resourceful'—they are adept at performing work without the best tools, sometimes by manufacturing their own special tools. Often, they are compelled to perform tasks without proper tools simply because of economic pressures. But once in a while, there are extreme situations, like the one illustrated in Example 3.4, wherein the company is not really equipped to perform repair using the most appropriate tools. At least in this case the company sought engineering approval to use different tools—whether such approval was a technical decision or a business decision is unknown.

Example 3.4: Composite repair with vacuum bags rather than autoclave

ASRS Report Number 442240

(Public document, edited for clarity)

> Mechanic was instructed to completely replace top, bottom skins and honeycomb core on ground spoiler. B-737-200 part number 65-46452 is an aluminum honeycomb sandwich—built and bonded together using molds and autoclave. This spoiler was rebuilt to a much lower strength than original, using only vacuum bag methods and no molds. After three attempts to pass inspection (bond voids were found), it is now in the parts system. The greater problem is that this 'new' process has been adopted for other damaged aluminum flight controls. [I] have tried to correct problem with internal hotline call without resolution. [I] have suggested that the on-site Boeing representative be asked if rebuilding flight controls using only vacuum bag methods is acceptable—says he is not an enforcement agency. There is a memo from engineering which changes the reading of structural repair manual instructions to allow vacuum bag only on curved surfaces. This saves several million dollars in molds and autoclave equipment, and gives us a leg up on the competition (that uses Autoclaves). Callback conversation with reporter revealed the following information: the reporter stated the procedure that engineering has allowed for flat composite surfaces, which is in conflict with the structural repair manual, has now been approved by the company engineering to use on curved surfaces. The reporter said this problem has been ongoing for a long time and has resulted in flight control surfaces from all aircraft owned by the company cycling through the shop with poor service time. The reporter said without the proper tooling a good repair cannot be accomplished.

Are You Getting Distracted from Your Task? Are You Being Asked to Attend to Another Maintenance Task While Performing Your Primary Maintenance Task?

Workplace distractions are among the leading causes of errors in aviation maintenance. Unlike an aircraft cockpit, aircraft maintenance environment is an open environment where it is common for a mechanic, especially in line maintenance, to be working on a job on one aircraft and be called on the radio to attend to another job on a different aircraft (sometimes several gates away) under very limited time. So, it is not surprising that people lose track of what they were doing and succumb to errors of omission: forget to torque a bolt, forget to safety wire something, forget to install an oil cap, forget to close-up the E&E bay door, forget to reset a circuit breaker, forget to remove landing gear locking pins, forget to remove their flashlight from the wheel-well, etc.

As you may have noticed in the list of typical errors associated with workplace distraction, the errors are errors of omission—forgetting to do something. Typically, a mechanic starts on one job and is called to work on another job prior to the completion of the first job. So, when he returns to the original job, it is likely that he has forgotten where he left off or the time pressure is very high and so the mechanic now has to rush through this job. The best way to minimize such errors is to devise a memory aid that will work not only for yourself but also for your workgroup so that it is easy to note the incomplete status of a given job. Forgetting to close E&E bay and forgetting to secure oil caps is so common that every mechanic must form a habit of not leaving these two items unsecured, ever. If you need to get away from the job for a moment try to attach a streamer if you have incomplete items or secure the E&E hatch or secure the oil cap before you leave. Failure to secure hatches and oil caps typically result in 'return to field' and therefore they are likely to cost not only the fuel burned in taxi, takeoff, and landing, but also the fuel that needs to be jettisoned to prevent overweight landing (see Example 3.5).

Example 3.5: A case of multiple interruptions leading to an error of omission

ASRS Report Number 431654

(Public document, edited for clarity)

> On the morning of mar/xa/99 aircraft xyz was scheduled for service check. As part of this check, aircraft hydraulics were pressurized and a visual inspection needed to be performed. I noticed #1 hydraulic system was low, so I serviced system and began checking wheel wells which required opening gear doors. After inspection, I closed main landing gear doors, and was called over to check on a contract aircraft. After returning to aircraft, I had to check on the fuel release which was late and had to address some cabin discrepancies, and then placard a lavatory inoperative. In the rush of getting the plane ready to leave, I forgot to secure the nose landing gear forward gear doors or to see if the other mechanic had secured them. After pushback, someone noticed the gear doors down, but we were unable to contact the flight crew prior to takeoff (they switched to tower frequency). Aircraft took off, felt vibration and dumped fuel to return to [field]. We inspected nose landing gear door and wheel well, which checked all right. Aircraft refueled and departed.

Are You Working Under Time Pressure? Is This A 'Rush' Job?

The line maintenance environment is particularly challenging because the window of opportunity to perform the requisite maintenance keeps getting narrower and narrower. At times, we find that the gate agents have already started boarding passengers before maintenance can clear the discrepancies. In such an environment, it is all a game of 'blame the delay'. Since a flight delay of over 15 minutes has to be reported, it has become an airline performance parameter. Consequently, there is tremendous pressure to have on-time departures. So, at the time of departure, there is a certain competition between gate agents and maintenance as to who is going to take the delay! In this process, mechanics are placed in a situation where they are troubleshooting a problem with no specific end in sight, the gate agents are calling every 5 minutes wanting to know whether or not they can board the aircraft, the flight crew wants all the discrepancies cleared and paperwork signed-off, and the maintenance managers/supervisors are managing multiple flights across the airport. In all, line maintenance is a highly demanding environment. One could argue that majority of the work in line maintenance is 'remove and replace' type rather than complicated troubleshooting or repair. It is these same removal and replacement type of jobs, however, that lend themselves to higher risk of forgetting to secure access panels, oil caps, air data computer lines, etc. Therefore, we suggest that you assume you are not going to get a second chance to secure a particular item or to clear all tools and follow these basic habits:

- *Do not leave any fitting finger-tight*: Torque all fittings to the appropriate torque the first time. You may not get a chance to return to this job and tighten it later. If it needs safety wiring, do it before you leave the scene.

- *Always secure access panels and hatches once the work is completed*: E&E bays are notorious for being left open. If you leave this panel open, the aircraft is guaranteed to return to field because it will not be able to pressurize. If that happens, you not only took the delay, but also lost significant revenue in wasted fuel (both consumed and jettisoned).

- *If you defer an item, check for status change*: Often with avionics items, if you defer something based on clearance from Maintenance Control, do not forget to ask whether the aircraft needs to be downgraded.

- *Always read every step that you sign-off*: In a rush, it is easy to sign-off a list of items without verifying each item in detail. That is when you risk signing-off items that may not have been cleared.

Are You Fatigued?

In the United States, one could work for 27 days without a break and take the last four days of the month off because the regulations are based on overall days worked, not continuous hours worked. The specific regulatory statement is as follows:

> 14CFR § 121.377 Maintenance and preventive maintenance personnel duty time limitations.
>
> Within the United States, each certificate holder (or person performing maintenance or preventive maintenance functions for it) shall relieve each person performing maintenance or preventive maintenance from duty for a period of at least 24 consecutive hours during any seven consecutive days, or the equivalent thereof within any one calendar month.

While the above regulation allows mechanics to work double shifts or longer shifts, it also sets them up for errors. When a mechanic has to work multiple 10-hour night shifts in a row, the likelihood of that person committing errors on the third and fourth night is very high.

As mentioned in Chapter 1, research by Dawson and Reid (1997) in fatigue has demonstrated that the degradation in decision-making skills due to lack of proper sleep for a 24-hour period is comparable to having a blood alcohol level of 0.10%–illegal to drive in most western countries. Also, Example 1.8 in Chapter 1 illustrates how a routine tire assembly task can be impacted by overwork and fatigue.

Work Management

Work management is a term used to describe the overall monitoring and execution of the work without taking undue risk. As such, you will need to not only plan your work so that you know what sub-tasks need to be accomplished and in what order, but you should also know the parts that are needed to accomplish your work and the appropriate technical reference materials required. As you continue to progress along the work path, you need to be cognizant of the time taken to accomplish your work with respect to the time available and actively assess the need to seek additional help. Also, sometimes people get frustrated with certain challenging tasks, especially troubleshooting of electronic components, which do not seem to be converging to a resolution.

Another important aspect of work management is management of distractions. It is recommended in most maintenance human factors courses that one should retreat three steps on a task card if one is distracted from the job. It is quite common for line mechanics to be working on the engine of one aircraft and be called upon to work on the landing gear of different aircraft on a distant concourse. By the time that mechanic returns to the original engine job, he may have forgotten where he was on the task card or worse yet, someone else may have 'finished-off' the job and released the aircraft. Therefore, it is important to placard your incomplete work and mark your spot on the task card so that you know exactly where you left off. Also, managers need to be more aware of the impact of such distractions.

Work Management Norms

The following norms have plagued the aviation maintenance industry for decades. The deliberate and conscious elimination of these norms is imperative. So, if nothing else, if you could make a deliberate effort to stop the following norms, you will make a significant contribution to aviation safety.

1. *People assume that parts that came off an airplane are correct and use them as reference to get the new parts*: Treat each disassembly as an opportunity to do a conformity check. Refer to the maintenance manual and the IPC to determine the correct part numbers and verify that each part that is put back on the airplane is correct, in accordance with the latest maintenance manual reference.

2. *Signing-off all items on a task card in one sitting*: The maintenance task cards are designed with the expectation that as the mechanic performs each task, he will sign it off on the card. In reality, most mechanics read the tasks on the card and put the task card in their pocket. Once they complete all the tasks (to the best of their memory), they return to the task card and sign-off all the items in one sitting. This practice has developed due to practical need to prevent grease and oil stains on the job cards. However, this practice has also led to some serious maintenance errors. This practice sets you up to sign-off tasks that you may not have performed.

3. *Leaving tasks incomplete, unattended, or untagged*: Workplace distractions are a reality. Whether the distractions are due to personal issues such as health problems in family, organizational issues such as impending layoffs, or operational issues such as someone needing help in resolving their tasks, the bottom-line is that they tend to affect the work performance. In our many years of visits to several maintenance facilities in the United States, we were truly impressed by the mechanics' ability to maintain their focus on quality in the face of personal as well as organizational issues. However, the operational issues that tug on the problem-solving and helping nature of these mechanics tend to trap them into making errors. Specific strategies need to be developed such that whenever a person is taken from one job to work on another, the incomplete tasks on the first job are clearly marked or tagged. If the task happens to involve high-risk (of forgetting) items such as oil caps or access panels, they should be secured or clearly marked prior to leaving that work area. Also, the lead or the supervisor needs to take responsibility for making sure that the mechanic returns to the original task in a reasonable time.

Example 3.6 illustrates a case from a confidential document. This case has been de-identified enough to protect the identity of the individual mechanic and the airline, but the relevant facts are presented unaltered. This case illustrates how multiple contributing factors such as incomplete or conflicting maintenance instructions, a job carried over multiple shifts, distractions during the job, and mounting schedule pressures can culminate in a maintenance error. Yet again, you will find that the mechanic who committed this error was in fact a very hardworking, diligent, and conscientious person. Given the multitude of causal factors, it could have happened to anyone. However, from a risk management perspective, the following pre-task factors should have alerted this mechanic to be extra cautious with this task:

- Confusing and incomplete maintenance instructions.
- Job carried over multiple shifts.
- Distractions caused by needing to go back and forth between the E&E bay and the rudder.
- Having to remove and replace the jumper wire multiple times.
- Working alone.
- Working from one set of instructions and signing off another.

Example 3.6: Multiple causal factors leading to a maintenance error

A Confidential Document

A B-737 aircraft was brought in for a Rudder PCU leak test in accordance with a task card. This job was first assigned on a night shift to a couple of new hire mechanics. These new hire mechanics working the task card encountered a problem with some of the later steps. They found that the Yaw Damper switch had been kicking off before the task card instructions direct that it be turned off. They interpreted this as a fault and consulted with their lead. Other work needed to be accomplished on this aircraft and they were re-assigned to accomplish these additional tasks. The difficulty with the test was passed on to the first shift's lead mechanic and his mechanics taking over the aircraft.

Before continuing with the task card, the first shift mechanics reviewed the wiring diagrams and determined that it was normal for the Yaw Damper switch to kick off when hydraulic pressure was removed. The first shift mechanics then realized that the steps in the task card were flawed. The first shift lead mechanic immediately consulted with engineering to address the conflict between the task card and the normal operation of the Yaw Damper system. Engineering later sent authorization to use a service letter to accomplish the PCU leak test.

The subject mechanic accepted second shift overtime and was assigned this aircraft, and was informed that the aircraft had a target release time of 20:00 hrs. The same lead that had worked the aircraft on first shift was staying over on the second shift. He did an excellent job briefing the subject mechanic on all of the above-mentioned events relating to the task card.

After reviewing the service letter and the task card, they began the test while also referring to the task card. The test voltmeter and amp clamp were still in place from where the previous crew left off. The subject noted that the meter readings they were getting were not correct. The service letter noted the normal range to be expected, but the task card did not. No meter hook-up instructions were provided in any of the maintenance instructions made available to the subject mechanic and his lead. The subject mechanic had performed similar leak checks numerous times and so he knew that a diagram with proper meter settings was available, but could not locate it in the instructions provided to him. His lead contacted several people/departments within the company, but could not locate the necessary diagram. During this hunt for the diagram and correct instructions, the frustration was building and the target release time was fast approaching. Finally, the subject mechanic contacted his friends at another station and had them fax a copy of the diagram.

Considering the mounting frustration and lack of help, they decided to bump the release time to 23:00 hrs. The subject mechanic reviewed all the maintenance instructions and determined that he would have to go back and forth between the tail compartment and the E&E bay for several sub-tasks along with several installations and removals of a jumper wire. Also, the service letter had instructions in a different order (they were in the correct order) than the task card (instructions were in the wrong order) that had to be signed-off. After satisfactory completion of the test, the directions in the service letter said to return aircraft to normal. The subject mechanic went over every step and returned all the valves and other items to normal. At that time, he missed removing the jumper wire from the Yaw Damper system.

The subject mechanic did not catch this vital error because he was working from two different documents under time and resource pressures as well as personal fatigue. Moreover, the two documents that the mechanic was using did not have the tasks in the same order, hence when he signed off the task card, he thought that he had indeed removed the jumper wire (as this step had to be carried out numerous times during the testing process).

After the aircraft was released, the crew reported difficulty in maintaining the wings level and had an unusual rudder trim correction. The aircraft returned to field.

Communication

Lack of interpersonal communication is perhaps the most significant problem when work is carried from one shift to another. Sometimes work that was designated to one shift gets started by the preceding shift with the noble intention to help, but could result in catastrophic consequences because the check-and-balances or 'barriers' of the maintenance system are not designed to handle such out-of-sequence tasks. See our previous volume (Patankar & Taylor, 2004 pp.176-179) for a detailed discussion of communication problems that resulted in the crash of a Continental Express flight in Eagle Lake, TX.

The Title 14 of the Code of Federal Regulations (14CFR) requires that interruption in workflow, whether due to shift change or multiple tasks being assigned to a particular person, be handled in accordance with a pre-approved protocol. The specific citation from 14CFR §121.369 (b)(9) is stated below, but similar requirement exists under §91.1427 (b) (9), §125.249(a) (vii), and §135.427 (b)(9).

14CFR § 121.369 (b)(9)

> Procedures to ensure that required inspections, other maintenance, preventive maintenance, and alterations that are not completed as a result of shift changes or similar work interruptions are properly completed before the aircraft is released to service.

These regulatory requirements address workflow interruptions due to shift change as well as distractions. So organizations operating under §91 (specifically, maintenance of fractional ownership airplanes), §121, §125, and §135 are required to have a specific process that is used by mechanics when their workflow is interrupted. The regulations do not say whether the process should be verbal, written, or combination; however, if we review some examples related to communication errors due to an ineffective shift-turnover process, it will be clear that perhaps the ideal process should employ a combination of verbal and written aspects. Also, in certain cases, it would be prudent for mechanics to actually show each other the work area and explain the status of the work. In reality, however, shift turnover is more like a telephone game: the outgoing mechanics convey the job status to their lead, the leads convey the job status to their foreman, the outgoing foreman conveys the job status to the incoming foreman, the incoming foreman conveys the job status to his leads, and finally the incoming leads convey the job status the incoming mechanics (Eiff, 1999). You can only imagine the level of fidelity (or lack of) in that type of turnover.

Example 3.7: Verbal turnover problem

ASRS Report Number 366822

(Public document, edited for clarity)

> I received a turnover for this aircraft which had a valve replaced for the left main landing gear. The verbal report I got was that the left landing gear control valve was replaced per maintenance manual. The actual valve that was replaced was the sequence valve per maintenance manual. Since I understood the control valve was replaced, I performed the return to service test which required the landing gear to be pinned and the landing gear handle cycled. The proper procedure for the valve that was actually replaced would have required the aircraft to be jacked and performed the landing gear retraction test (ref 32-32-0) and manual extension test (ref 32-34-0) required by airworthiness directive 79-04-01. The aircraft was stopped in xyz so that the correct retraction and manual extension tests could be performed. The problem was discovered by maintenance control center. This problem occurred because of a verbal turnover in which the valve was called by the wrong name.

Chapter Summary

In this chapter, we discussed three broad categories: preparation, work management, and communication. Each of these categories has certain elements that need to be addressed by mechanics and supervisors prior to starting a given maintenance task and also while performing the task. Some issues such as individual preparation—intellectual and physical capabilities and availability of resources—are fairly obvious. Work management and communication, on the other hand, require some practice. Work management skills are critical in proactively identifying high-risk tasks such as the example in Example 3.6. Communication skills can be formalized through a required pre- and post-task briefing among mechanics and their leads and/or outgoing mechanics and incoming mechanics.

Review Questions

1. Describe some of the situations wherein you did not have proper training or equipment to perform a maintenance task assigned to you.
2. What specific strategies would you use to better manage complex maintenance tasks—those that spread across shifts or involve convoluted/confusing troubleshooting procedures?
3. Describe the most effective shift-turnover you have either given or received. Ideally, what do you expect to communicate in this process?

Chapter 4

Post-Task Analysis

Instructional Objectives

Upon completing this chapter, you should be able to accomplish the following:

1. Describe multiple ways to practice assertiveness in your professional work.
2. Describe examples of lack of assertiveness, poor teamwork, and lack of integrity in the aviation maintenance environment.
3. Describe the use of pre- and post-task analysis scorecards to track individual as well as organizational progress in improving safety practices.
4. Explain the difference between rule-based safety and risk-based safety.
5. Describe the application of the Hawkins-Ashby model to practice risk-based safety.

Introduction

In this chapter, we will discuss three key issues in post-task analysis: assertiveness, teamwork, and integrity. Post-task analysis is not really limited to a particular task asking you whether or not you closed a panel or removed the landing gear pin, but it is a moment for introspection about your work habits. While pre-task analysis alerts you to consider some of the latent failures in your own work practices as well as in your organization's culture, the post-task analysis will encourage you to consider how you could improve your work practices so as to keep going beyond your normal call of duty and effect long-term systemic changes in your organization and your profession.

Assertiveness

Many human factors awareness programs emphasize the importance of assertiveness in terms of the need to speak-up. As a direct result of such programs, we have documented an increase in the opinion about assertiveness, meaning that increased number of participants believe that they should speak-up. Raising the overall awareness regarding the importance of assertiveness is certainly important, but too often it tends to be limited to speaking-up.

In our view, assertiveness involves active listening, speaking-up, and acting on concerns that are voiced. Active listening is a skill that enables people to listen to a message attentively, consider both aural signals as well as body language, and confirm that they have understood the message by restating the message to the original speaker. For pilots and air traffic controllers, this is a well-polished standard practice. Pilots are quite fluent at receiving verbal instructions, noting the most critical aspects of those instructions, and repeating them to the controller's satisfaction. However, even pilots may not practice this skill consistently outside their cockpit.

In the course of our research, we have visited numerous maintenance facilities in the United States. As we observed the pre-shift briefings done by a foreman or a supervisor, we noticed that for the most part, these were one-way news bulletin style communications that were barely audible at the back of the large break room or cafeteria where they were typically held. Most attending mechanics tend to ignore such briefings because the messages in the briefings are generally not specifically interesting to any one mechanic or workgroup. Even when a scaled-down version of that meeting is held between a lead mechanic and his three to four mechanics, it tends to be a one-way communication: the lead announces how the jobs are allocated and hands out the job/task cards, no questions or discussion.

Ideally, there should be both pre- and post-shift meetings that are highly interactive—every participant is given the opportunity to speak-up and specific work-related issues are discussed. Typical hurdles in accomplishing such meetings are as follows:

- The company's parking garage cannot accommodate both incoming and outgoing shifts for any overlapping periods.
- There is no time to hold a pre- and post-shift briefing.
- The mechanics and their supervisors are not trained to handle an interactive meeting.
- In large companies, particularly with the rotating days off system, leads or foremen have no idea who they are going to work with on

a particular shift. Consequently, they are not prepared to handle variations in skill set. Having a pre- and post-shift briefing would not be as effective if you have to work with a revolving set of people. In one maintenance facility, a lead said to us, 'Give me the worst people you got, but let there be some consistency so that I can train them!'

The Concept Alignment Process (CAP) as an Assertiveness Tool

The Concept Alignment Process was first adopted by a corporate aviation department's flight crew in 1995 and subsequently (in 1998) customized by the maintenance department to suit their needs (Patankar & Taylor, 1999b). The CAP is different from most of the Maintenance Resource Management (MRM) programs because it focuses on a behavioral change rather than an attitudinal change. The organization need not change everyone's safety attitude before expecting a change in behavior. The CAP forces all employees to change their behavior and follow a prearranged process. Therefore, it does not suffer from the limitations of the first three generations of MRM programs (c.f. Taylor, Robertson, Peck, & Stelly, 1993; Taylor, 1995, Taylor, Robertson, & Choi, 1997; and Taylor & Christensen, 1998).

The CAP is a simple communication protocol that provides means for all the individuals to share information. At the heart of this protocol is the *concept*. A concept is defined as an idea, remark, or an observation that is stated by one person and is either affirmed or challenged by the co-worker. If a difference between the points of view is stated, it is the team's responsibility to seek validation for that concept from an independent third source. If one concept can be validated and one cannot, the validated concept shall become the working concept. If both can be validated, the choice of which one becomes the working concept is up to the team coordinator (designated prior to starting the task). If neither concept can be validated, the most conservative of the two is chosen. Once a working concept is agreed upon, it is further scrutinized using a predefined judgment process. Often in this process, the technicians, management, and flight crew research the cause of the discrepancy in the concepts and recommend appropriate changes. Changes have been made in operating policies and procedures, maintenance manuals, and other documentation as a direct result of this process. (Patankar & Taylor, 1999b).

Patankar and Taylor (1999b) further reported that not all of the technicians in this organization practiced the CAP to the same extent. For example, some of them understood the basic protocol, but hesitated to challenge another person's concept or to seek validation. Only a couple of individuals were observed to be practicing the CAP consistently and to its full potential (challenging concepts, seeking validation, identifying causes for ambiguity in information, and implementing appropriate structural/procedural changes so that the ambiguities are minimized). Patankar and Taylor also observed that as the skeptics used the process, they understood it more clearly, and as their success in effecting organizational changes grew, their faith in the process grew. Gradually, they were becoming believers. Hence, this company was able to cause an attitudinal change through behavioral change, rather than the other way around.

In this case, the technicians believed that they overcame problems in active listening such as 'listening but not hearing' and recognized that they were *rationalizing* old procedures rather than *validating* them by not practicing assertiveness through the CAP.

Example 4.1 below illustrates a case from the above-mentioned corporate aviation department where pre- and post-shift briefings are required. The level of assertiveness and interaction that took place in these meetings are a direct result of the CAP implemented at that company.

Example 4.1: Landing gear fairing bolt case

Aircraft A's main landing gear was being lubricated during an inspection when it was discovered that the aft trunion on both sides would not take grease. Maintenance tried several attempts unsuccessfully to get the trunion to accept grease. The fact that the trunions did not take grease is a concept. As such, it must be validated; if it can be validated, it should be acted upon. In this case, the concept was clear and could be easily validated by observing that the trunions did not accept grease. Next, the technicians had to proceed to the judgment phase wherein they had to choose a course of action. The maintenance manual was clear regarding the course of action (If the bearing will not accept grease then replace the bearing); therefore, they decided to change the bearings.

The parts were installed in two shifts. Starting with the right main landing gear (MLG), which was completed by day shift. Swing shift continued with the left MLG following the shift briefing. As the job was being finished a problem was found with the MLG cylinder-retaining nut. This nut required a torque of (21.39 to 30.24 ft.-lbs.). The nut stripped as it was torqued beyond 10 ft.-lbs. Removal of this nut revealed that the bolt was also damaged. This was reported in the maintenance turnover via voice mail for maintenance the next day.

It was discovered during the next day's shift briefing, that the previous day when the right MLG was assembled, the day shift technician had made an individual judgment when the MLG cylinder nut was installed. He torqued the nut just below the rated torque (20 ft.-lbs. Instead of the minimum 21.39 ft.-lbs.) to preclude the possibility of stripping it. His comment to the maintenance manager was that after looking at the installation, he 'felt it would be fine just below rated torque with the cotter pin installed'. This decision was based on experience with similar installations and echoes technical trainers' customary admonition, 'Be careful not to over torque [it]'. From a CAP perspective, the concept—nut does not take rated torque—was never stated. The subsequent course of action—to torque the nut below the minimum recommended torque—was never validated using a third-party source. When the technician told his manager about his decision, the manager understood how the technician had come to the decision to under-torque the nut, but there was no validation based on an independent third-party. The technicians recognized that this should have been the opportunity to 'challenge the concept'. Replacement hardware was installed on both MLG cylinder-retaining fittings and torqued to specifications.

The technicians viewed this case as an issue of miscommunication between the day-shift technician who under-torqued the nut and the swing-shift technician who stripped the threads. The training coordinator, who was a member of the aviation department, ruled differently. According to him, the fact that the day-shift technician thought that it was acceptable to under-torque (below the minimum requirements) was a violation of the company's policy. According to him, if the swing-shift technician had also under-torqued the nut, nobody would have known about it. So, rather than a case of miscommunication, it is a case of misalignment of the core values—by torquing outside the specified limits, the company's policy was violated.

Teamwork

In maintenance, it may appear as though mechanics are functioning as a team. In reality, however, they are functioning more like a group with each member having a set of predefined tasks to perform. Someone higher in the management or in the planning department knows how all of these tasks come together and accomplishes the collective goal, but as far as the individual mechanic who receives a set of job cards at the beginning of his shift, there is rarely any awareness of the collective goals. That is not to say that the mechanic does not know whether or not he is working on a 'C' check, but that he may not know how his work might impact another mechanic's work. For the most part, when mechanics receive their job cards, they pursue their assigned tasks almost like independent workers.

A team is a group of individuals coming together to achieve a common goal. As such, they must fully *understand the common goal*; *communicate* their individual knowledge, ideas, and concerns; and *trust* that each team-member will do their best to execute the assigned responsibilities. The extent to which these three needs are satisfied will influence the overall effectiveness of the team.

Another way of looking at teamwork is to consider the human reliability example discussed by Patankar and Taylor (2004). In accordance with that example, systemic reliability is maintained by providing redundancy in either number or function. As such, an effective team will be capable of absorbing certain types of errors by individual members because they will have an effective defense mechanism such as cross-checking each other's work, briefing each other of the potential errors in a particular job, and finally simply doing a team-member's task if that member is unable to perform that particular role/task.

In reality, we find that some mechanics who are supposedly working in a team with engineers, inspectors, and management, rarely communicate with each other. Consider the example in Example 4.2. In that case, we have several individuals involved in a job, but no one keeping a global track of what was going on or even practicing closed-loop communication between as well as across the professional groups—engineers not talking with other engineers, mechanics not seeking validation from engineers prior to executing the repair, and inspectors not talking with other inspectors.

Example 4.2: Lack of communication and teamwork

ASRS Report Number 460228

(Public document, edited for clarity)

> I was called to join a maintenance team, in the effort to repair structure damage to the forward pressure bulkhead of a McDonnell Douglas DC-10-10. The aircraft was damaged in zzz airport inside one of the maintenance hangars. Another aircraft being moved struck the radome with the right hand elevator, causing damage to the radome and punching a small hole about 6 inches by 8 inches in the forward pressure bulkhead. The damage was removed, and the engineer was called to evaluate the removed damage, and came up with a repair procedure. The first instruction was to have a local aircraft inspector 'non destructive test' the area for cracks and defects. No defects were found. The engineer gave us a verbal repair proc to be followed by written documentation of the repair. Skin was to be repaired per the structural repair manual (manufacturer's maintenance repair manual) and ATA Chapter 53-20-00. One problem occurred during this portion of the skin repair—the fasteners that were

called out to be used, per structural repair manual, would interfere with fasteners installed previously on the forward bulkhead repair. This problem was brought to the attention of the engineer. The engineer's evaluation was to install a substitute type of fastener, which would deviate from the structural repair manual. This was a verbal instruction, and was assumed it would be annotated in the engineering variation authorization, written by the engineer. On completion of the left-hand skin repair, a second engineer from the engineering department appeared to look at the completed repairs on the left-hand fuselage. The engineer appeared to take measurements. The aircraft maintenance technicians asked the second engineer if everything looked to be per specifications. The engineer replied 'yes, it looks good'. The company aircraft operating procedures require the aircraft maintenance technician to sell the work done to a buyback inspector local inspection was used. The question was asked (between the aircraft maintenance technicians) if the inspection department had qualified people trained in major structural repair and procedure for buyback. No answer was given. We assumed there was. The inspection department sent one of their inspectors to review the repair. The inspection department found a number of fasteners (rivets) that had been over bucked ('in his opinion'). The inspector requested to have them changed, and to call him back when finished. The fasteners were changed by the aircraft maintenance technicians. The inspection dept was called to send out the same inspector back to the aircraft, for buyback. We were informed the inspector had gone home. Another inspector was sent over. The aircraft maintenance technicians had informed him of the first inspector's find. Approximately 10 fasteners had been replaced. The inspector found no defect, and approved the repair. Still, the aircraft was not released to service, until all of the cables, insulation, brackets, etc (removed for access) were reinstalled, and operationally checked. Once completed, the aircraft was released for service. I was informed by one of the supervisors that the aircraft I was involved in repairing had been grounded in xyz airport. The reason was the l-hand skin repair was done improperly. Someone had informed our quality assurance department that the repair was not installed per the structural repair manual document (53-20-00 class 2 type). I do not have any details to what the corrective action taken was, only that the aircraft had been returned to service. An investigation is being conducted by our quality assurance department at this time. Human performance considerations: the fact that the first engineer was called back from his vacation to perform a repair package should have never taken place. I believe he was not in the right frame of mind to perform this job. To have another engineer try and finalize paperwork for someone else, after the repair was completed, was also an error. Our assumption that the engineer would document the changes was an error as well. With the holiday approaching, I'm sure it played a certain amount of effectiveness in state of mind, for all those involved. The improper training or [lack of] familiarity with major structural repairs

> within the inspection department should have been addressed by the buyback inspector or the inspection department supervisor. The pressure from the upper management to return the aircraft to revenue service played a big role.

It is interesting to note that the author of the report presented in Example 4.2 was quite aware of multiple causal factors leading to this event; however, this event could have been easily avoided with basic teamwork. If the engineer who verbally approved the substitution of the rivets had followed-up with a written authorization, the repair would have been legal. For some reason, the mechanics lacked the assertiveness to express their need to have the written authorization prior to performing the work. The management, on the other hand, seemed to be sensitive to the schedule pressure, but did not do much to facilitate communication between the various people involved in this repair. Ultimately, everyone had somewhat different goals: the management wanted to get the airplane out on time, the engineer wanted to avoid documenting his prescribed repair, and the mechanics wanted to get the job over with. Proper teamwork would have ensured that all the paperwork for this job was in order and that the aircraft could be legally released for service.

Teamwork as a Means to Improve Ethical Decision-making

According to Kohlberg's theory of moral development (cited by Beabout & Wennemann, 1994), an individual's decisions can be categorized as follows:

1. *Level 1: Self-interest*: Level-1 decision-makers base the rationale for their decisions on pain aversion. Therefore, they make their decisions either to comply with a particular regulation to avoid legal prosecution or to prevent their supervisor's punitive treatment.

2. *Level 2: Conformity to One's Society*: Level-2 decision-makers are motivated by their need to conform to their society—basically, they are simply trying to fit in.

3. *Level 3: The Principle of Respect*: Level-3 decision-makers respond based on the principle of respect to all and give full consideration to the moral responsibility placed in him by the profession/society and his own need for similar consideration by others if the roles were reversed.

Considering the above three levels, professional as well as labor organizations could promote a high safety standard that its members must abide by in order to be accepted in the organization. In the formative years of their career, mechanics are just as moldable as any other young professionals. It is a professional and moral obligation of the more experienced individuals to groom the younger generation such that they become the future safety champions.

Integrity

Integrity (or lack thereof) is arguably the fastest growing concern in aviation maintenance. On the very basic level, integrity is consistency between words and action. When a person with high integrity signs-off a maintenance task as complete, it is indeed complete. If a person has low integrity, the task may or may not actually have been executed: just because he signed-off an aircraft as airworthy does not mean that he really did his best effort in ensuring that the aircraft was airworthy.

Relationship Between Trust and Integrity

Based on our survey research, we reported that up to one-third of the mechanics in the United States do not trust that their managers will act in the interest of safety (Taylor & Thomas, 2003b; Patankar & Taylor, 2004; Patankar, Taylor, & Goglia, 2002). When we presented these results at several conferences, most mechanics thought that those results were under-rated. Our survey asked the respondents to rate the following items on a scale of 1 to 5 (1 = Strongly disagree, 2 = Disagree, 3 =Neutral, 4 = Agree, 5 = Strongly Agree):

- My supervisor can be trusted
- My safety ideas would be acted on if reported to supervisor
- My supervisor protects confidential information
- I know proper channels to report safety issues.

As you may have observed, the issue of trust is a matter of relationship between two or more individuals. The questions asked in the survey seek to quantify the degree to which one individual trusts another. Integrity, on the other hand, is a matter of individual ethic or moral conduct. A person is believed to have integrity when the following observable behaviors are consistent (derived from Beabout & Wennemann, 1994):

- The person's actions are consistent with his/her words—'can talk the talk and walk the walk'.
- The person is prompt in accepting responsibility for his/her mistakes and takes concrete steps to minimize the recurrence of similar mistakes.
- The person keeps his/her promises.
- The person is acutely sensitive to professional and social responsibilities.

So, the issues of trust and integrity are inter-related. When there is integrity within an individual, other people are likely to trust that individual.

Professional Integrity

The primary professional responsibility of an FAA-certificated aircraft mechanic or a CAA-licensed aircraft maintenance engineer is to evaluate the airworthiness of an aircraft. Every time a mechanic or an AME signs-off the airworthiness release of an aircraft, he is attesting, to the best of his knowledge and abilities, to the mechanical integrity and legal compliance of that aircraft.

As a result of some recent aviation accidents, a phenomenon called *pencil-whipping*, a term used to describe the practice of falsely signing-off maintenance tasks as completed, has surfaced to public attention. The issue of intent behind such practice is a matter of much debate. While there are some atrocious examples of intentionally falsifying airworthiness releases, there are likely to be several more examples of unintentional signing-off of routine maintenance tasks because some mechanics have gotten into a habit of signing-off all the subtasks on a task card in one sitting.

Does improved reliability engender risk-taking behavior?

According to a Boeing report (Boeing, 2003) on hull-loss accidents, a review of the ten-year (ten years since the introduction of the aircraft) accident rate indicates the distribution of accident rates per million departures as follows:

- First generation aircraft (707/720, DC-8, Caravelle, etc.) = 27.4
- Second generation aircraft (727, DC-9, 737-100/200, etc.) = 2.9
- Early widebody aircraft (747-100/200/300/SP, DC-10, L-1011, etc.) = 5.2
- Current aircraft (MD-80/90, 767, 757, A300-600, 777, etc.) = 1.5.

Obviously, the change in the accident rate from the first to subsequent generations of aircraft is significant. Some argue that this initial decline in the accident rate is attributable mainly to the design improvements by manufacturers such as the Boeing Company, and we are at a point where safety benefits from design improvements have saturated. Consequently, further reduction in accident rate is more likely to come from improved understanding of human factors and socio-technical issues (as discussed in Chapter 2).

Since 1989, human factors issues in maintenance have been studied in great detail. As a result of these studies, we now have documented evidence about maintenance errors and some of the causal factors associated with those errors. One question that comes to mind after reviewing the data on maintenance errors, particularly the pencil-whipping aspect, is whether or not the risk-taking tendency among mechanics and managers increased since the 1959-1979 period wherein there was a dramatic decline in the accident rate of commercial airplanes worldwide (Boeing, 2003). Since we do not have any hard data about the maintenance errors and the extant safety culture during the 1959-79 period, one can only hypothesize based on anecdotal reports from experienced mechanics claiming that, in the old days, the aircraft flew with fewer deferred maintenance items, the mechanics were better trained/skilled at their jobs, most people had a stronger work ethic, and profit was not as strong a motive because the industry was regulated and airlines competed on service quality rather than air fares. Again, it is not clear whether the above listed factors correlate better with deregulation of the industry, enhancements in technical reliability, or lack of influx of aircraft mechanics from the military. Nonetheless, the current reality is that production pressures are limiting the scope of inspections and consequently affecting the quality of maintenance/safety. Example 4.3 illustrates one such example.

Example 4.3: Production pressures limiting the scope of inspections

ASRS Report Number 458840

(Public document, edited for clarity)

> On Nov/xa/99, I was assigned an A-check on the line here in ZZZ. My job was the routine check and inspection of #1 and #3 engines. During the course of the inspection, I found items that needed to be looked at further. So I wrote them up in the logbook. After returning to work, I was told by my management that this was not a 'Hangar A-check' and that some of my write-ups were 'not valid' even though the suspect parts were replaced. I have not received a warning letter in my personal file for not understanding the scope of line inspections nor were they (the write-up

> areas) included in the A-check. I have copies of the write-ups and a copy of the air carrier A-check. I was taught in A&P school that if you had a question about something, write it up and do the corrective action. My management team here in ZZZ has created a hostile work environment and wants mechanics to 'pencil whip' jobs here in ZZZ and if you do not toe the line, you'll be written up for doing your job. Callback conversation with the reporter revealed the following information: The reporter stated a number of reports were written-up on discrepancies found on #1 and #3 engines during the A-check in question. The reporter said the major problems were found where the main engine fuel control rod ends were worn beyond maintenance manual limits and required replacement. The reporter stated that the local maintenance management wants to limit the scope of the check and avoid write-ups. The reporter said that the FAA Primary Maintenance Inspector has been checking the A-check paperwork recently but no findings or actions have been taken.

Beyond Regulatory Compliance

The notion of risk-based safety rather than rule-based safety is gaining momentum among safety researchers. David Marx argues that the current rule-based environment is not likely to enable significant reductions in aviation accidents. His basic argument is that it is very difficult to make effective rules for every situation, and even if one is successful in creating such rules, it is likely that over a period of time, these rules will either lead to an accident because someone followed the rule to the letter or it will be very difficult to update the rules in order to reflect changes in technology or corporate practices. Furthermore, the reason people develop norms or work-arounds is that they find the existing rules to be ineffective. Marx proposes a disciplinary policy that is based on 'recklessness'—which he defines as intentional violation of an accepted professional practice, organizational policy, or federal regulation, so as to endanger the safety of flight. Marx wants the aviation industry to acknowledge that certain rules may have to be violated at times, but also to emphasize the responsibility for making decisions based on the risk of the consequences associated with those decisions. (Patankar, 2003a).

Are We Saturated In The Level Of Safety?

Many industry professionals tend to attribute the current level of safety in aviation largely to design improvements, including systemic redundancies and component reliability. These improvements are robust enough to withstand compromises to operational as well as maintenance expectations. In order to enhance flight safety any further, the aviation system must consider human reliability issues and re-design the support infrastructure to tolerate individual human failure without compromising systemic failure. These ideas were presented by us in greater detail in our previous volume (Patankar & Taylor, 2004). In this section, we would like to connect those ideas with the notions of risk-based discipline and safety ethics.

The ethical dilemma used by Kohlberg (1984) to study the different levels of moral development involves violation of a rule to save a life. His study concludes that those subjects who thought that it was morally obligatory to violate the rule 'not to steal' in order to save a life, did so because the intent, motive, and circumstances involved in that case provided an overwhelming support to the first principle of ethics—mutual respect—albeit at the cost of rule/regulatory violation. Kohlberg further claims that the subjects who supported the above judgment, were more 'mature' in their moral development because they did not decide to violate the rule in order to avoid some sort of pain or in order to fit in a social group; they did it because they wanted to show respect to the person whose life they wanted to save.

Now, let us connect the above scenario with David Marx's risk-based model of disciplinary action/policy. According to Marx, the above type of scenario should be accepted into an ASRS or an ASAP process because although the subject violated a rule, the societal benefit of that violation far out-weighed the risk. Furthermore, by accepting such a report under the ASRS or the ASAP system, the circumstances leading to the violation could be understood and possibly, if warranted, the basic rule could be changed.

Marx offers another perspective to view the above scenario. Since there are so many regulations in aviation, it is unrealistic to expect everyone to comply with all the regulations all the time. Also, it is normal for aviation professionals to develop workarounds in their routine activities because of a wide variety of reasons including the following: (a) lack of availability of people, approved tools, or parts, (b) lack of feasibility to provide appropriate and timely training, (c) impractical or unworkable maintenance procedures, and (d) mounting schedule pressures (Patankar, 2003a).

According to Marx, the ASAP or ASRS cases should be judged based on whether or not they involve reckless behavior—one that intentionally endangers the safety of flight. Intentional disregard to safety is same as recklessness, but intentional rule violation is not necessarily recklessness.

The concept of risk-based decision-making rests on another concept—the level of information uncertainty. In order to make a safe decision, one has to rely on a certain level of knowledge of the risks involved. If this knowledge is corrupted or ambiguous, the ability to make safe decisions is impacted. Furthermore, it is very likely that the information as well as its level of uncertainty will change over time. Inability to effectively monitor the continued viability of a decision may also impact risk-based decision-making. That brings us back to Level-1 decision-makers versus level-3 decision-makers discussed earlier in this chapter.

In the event of high uncertainty about the impact of a particular decision, one should make rule-based decisions (Level-1). In the event of low uncertainty regarding the impact of a particular decision, one may make a risk-based decision (Level-3).

What happens when a mechanic thinks that he has sufficient level of certainty to make a risk-based decision, but subsequently discovers that the level of uncertainty was much higher? From an ethical perspective, Beabout and Wennemann (1994) suggest that the individual should be willing to accept the consequences of that decision. In the case of aircraft maintenance, the consequences could be not only legal prosecution, but also loss of life or property.

Field personnel are typically expected to make rule-based decisions and managers are typically expected to make risk-based decisions. Recklessness could occur at both levels; however, the impact of such actions is likely to be different. For example, assigning personnel to perform tasks for which they are not adequately trained is recklessness on the part of the management. However, due to the Federal Aviation Regulations, the mechanic performing the task is *always* held accountable, regardless of the circumstances.

A Study of Procedural Compliance Among Flight Students

It would be worthwhile to digress for a few moments to review a couple of studies done by Patankar and Northam (2003a, 2003b) about procedural compliance among flight students because flight training activities in the United States may be performed under one of two types of FAA approvals: 14CFR §61 or 14CFR §141. Compared to §141, §61 has less stringent requirements concerning periodic evaluation of the flight students as well as those concerning specific lesson plans. Therefore, §61 is less driven by rigid rules compared to §141.

The first study (Patankar & Northam, 2003a) focused on the comparison of attitudinal differences and the second study (Patankar & Northam, 2003b) focused on behavioral differences. Results of these questionnaire-based studies in structured training environments support the claim that 100% procedural compliance is impossible under normal operating conditions. Therefore, an alternate safety philosophy—based on one's personal values and risk tolerance—needs to be applied.

Classic safety studies have been retrospective in their attempts to seek causal factors leading to specific accidents/incidents. In the process of identifying such factors, investigators often tend to (in some cases, they are required to) cite specific regulatory requirements that were violated. Consequently, procedural and/or regulatory compliance has become the safety standard by default. Although many aviation organizations tend to strive toward higher safety standards, factors such as economic hardships and corporate acquisitions and mergers make it very difficult for some of the best-intentioned organizations to 'remain legal'. Then, it is just a matter of time until the violations are either detected by the regulators or they result in a mishap (Phillips, 2002; CNN, 2000).

During the years 1998 through 2000, there were 1555 cases of enforcement actions by the Federal Aviation Administration against aircraft mechanics employed by either an airline or a repair station (Patankar, 2002). Similar studies in the flight operations domain are not known. However, recent CRM research, reporting 'normal behavior' of flight crews, concluded that the differences in threat recognition and mitigation among the different flight crews were significant: some crews encountered less threats than others, some were better managers of threats, and some were quite poor in their recognition/management of threats (Helmreich, In Press; Helmreich, Klinect, & Wilhelm, 2001; Sexton & Klinect, 2001). Considering both the volume of regulatory-violation cases against aircraft mechanics as well as the limited threat detection and mitigation abilities of certain flight crews, one could conclude that under normal operational conditions, both flight crews as well as mechanics are susceptible to procedural violations—deliberate non-compliance with known organizational or regulatory requirements.

Helmreich et al. (2001) discovered that in an attempt to understand why one set of flight crew members would be more compliant with the standard operating procedures or why one set would be better at managing threats, the classic CRM factors such as communication, assertiveness, teamwork, etc. resurface. Also, cultural differences due to differences in nationality and organizational affiliation play a significant role in the overall performance of the different flight crews (c.f. Helmreich & Merritt, 1998).

In spite of such extensive longitudinal studies about pilot attitudes and behaviors in the airline community, such studies in the flight-training domain—the primary building block of pilot attitudes and behaviors—have been limited.

In a study of pilot attitudes and behaviors among flight students, Patankar and Northam (2003a) reported that about 10-20% of the flight students are not consistent in their compliance of required procedures. Thus, these students may be taking undue risk. If up to 20% of the flight students are likely to violate required procedures while in a structured training environment, it could be postulated that such violation is higher in the less-structured environment outside the flight school.

Method for the Behavioral Study: Patankar and Northam (2003a) used a survey questionnaire, called Flight Instruction Safety Culture Questionnaire (FISCQ) to measure safety attitudes of individual pilots as well as the organizational safety culture among 28 flight-training organizations. The FISCQ was filled-out by 100 pilots (student pilots, private pilots, and commercial pilots) undergoing flight training. In this section, we present the analysis of items 5, 7, and 14 from the FISCQ that were chosen because they were directly related to procedural issues. These items are presented in Table 4.1. A simple frequency analysis of the responses to the items listed in Table 4.1 was performed.

Table 4.1: Attitudinal items corresponding to procedural compliance

Item Number	Description
5	I work better when operational procedures are flexible
7	The flight training organization's rules should not be broken—even when the employee/student thinks that it is in the organization's best interest
14	Written procedures are necessary for all in-flight situations

A one-way Analysis of Variance (ANOVA) test was performed on each response item to determine whether or not the differences in the mean scores for each certificate type were statistically significant ($\alpha = 0.05$). An independent samples *t*-test was performed to determine whether or not the difference in the mean scores between the two types of flight-training schools was statistically significant ($\alpha = 0.05$).

Results: The one-way ANOVA test for each of the three response items (5, 7, and 14) indicated that there were no statistically significant differences across the three pilot groups, $p > .05$. Similarly, the independent samples *t*-test for the above-listed items did not yield a statistically significant difference between the responses from the two types of flight-training schools. Hence, one could assume that the sample was somewhat homogenous in their responses to all three items that were analyzed. The item-wise results are presented below. See Figure 4.1 for a summary illustration.

- *Item 5: 'I work better when operational procedures are flexible'.* On a rating scale of 1-5 (1=Strongly Disagree and 5=Strongly Agree), 47% of all the respondents either agreed or strongly agreed with this statement; 26% either disagreed or strongly disagreed; and 27% remained neutral.

- *Item 7: 'The flight-training organization's rules should not be broken—even when the employee/student thinks that it is in the organization's best interest'.* On a rating scale of 1-5, 54% of all the respondents either agreed or strongly agreed with this statement; 16% either disagreed or strongly disagreed; and 29% remained neutral.

- *Item 14: 'Written procedures are necessary for all in-flight situations'.* On a rating scale of 1-5, 37% of all the respondents either agreed or strongly agreed with this statement; 31% either disagreed or strongly disagreed; and 32% remained neutral.

Discussion: Item 5 of the FISCQ seeks responses regarding the participant's preference for flexibility in operational procedures; item 7 seeks responses regarding the basic belief that procedures should be followed at all times; and item 14 seeks responses regarding the individual's need for procedures.

The above results indicate that in this sample, 47% of the respondents prefer operational flexibility, 16% of the respondents are likely to violate the organization's rules (especially if they think that such violation is in the best interest of the organization), and 31% of the respondents do not seem to think that written procedures are necessary for all in-flight situations. From a previous study by Patankar and Northam (2003a) it is known that up to about 20% of the student population is not likely to be consistent in their adherence to required procedures. Consequently, one could conclude that at least one-third of the pilots in this sample have a poor attitude toward required procedures and up to about one-fifth of the sample is not likely to be consistent in their procedural compliance.

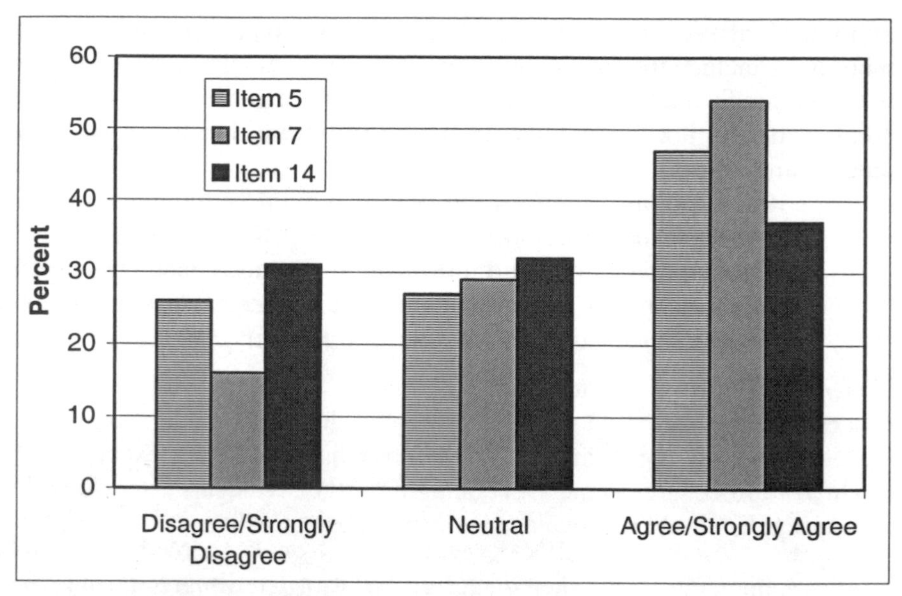

Figure 4.1: Summary of item-wise analysis

Value-based Safety: A Hawkins-Ashby Model. Patankar and Taylor (2004) presented a unique match between the SHEL(L) model developed by Hawkins (1987) and the *Law of Requisite Variety* developed by Ashby (1956). This model was reviewed in the present volume in Chapter 3.

Patankar and Northam (2003a) reported that 10-20% of their subject pilots were inconsistent in their compliance with the required procedures. If these students were taught to think and decide in accordance with the Hawkins-Ashby model presented earlier, they might realize that by deliberately not following some of the required procedures, they have placed their aircraft ('Hardware' according to the SHEL-L model) outside the system. Thus, one or more of the other components must compensate for such a change. On one specific survey item, about 8% of the respondents indicated that they may have departed on a cross-country flight without the required amount of oil in their engine. Such a violation clearly places the hardware out of the system and makes it imperative that the remaining components make the requisite changes (examples of such changes include restricting to local flights instead of cross-country flights, being more vigilant about the oil temperature and pressure, and being more vigilant about an emergency landing site).

Those students interested in a more proactive application of the Hawkins-Ashby model may want to consider developing a list of personal minimums that aid them in making prudent decisions and ensure that safety of flight is always the primary consideration. Examples of such personal minimums include the following:

1. 'If I do not touch down by the first taxiway, I will execute a go-around'.
2. 'If I am not airborne by a pre-specified point on the runway, I will abort the takeoff'.
3. 'When in doubt about any information—weather report, air traffic control clearance, aircraft performance data, or airspace issues—I will seek clarification from a reliable third party'.

Conclusion: Given the tendency of student pilots to violate required procedures, a value-based model is presented. In such a model, emphasis is placed on raising the awareness of the consequences of not executing the required procedures and encouraging the students to develop appropriate/adequate risk management strategies.

Implications of the Flight Training Study to the Maintenance Environment

In the maintenance environment, although we could have maintenance organizations certificated under either §121 (Airline) or §145 (Repair Station), the procedural compliance habits between these types of organizations cannot be compared because 90% of the mechanics in §121 organizations are FAA-certificated compared to only 47% of the mechanics in §145 organizations (Goldsby, 1996). Nonetheless, one important point to note that Patankar and Northam (2000b) recommend that pilots develop their 'personal minimums' that are in fact personal habits beyond regulatory requirements. These habits are also based on the pilot's understanding of the risks involved. Similarly, mechanics could develop their personal minimums that they would adhere to when faced with difficult decisions. Some sample personal minimums for mechanics are as follows:

1. I will keep track of my training records so that I will be able to alert my supervisor when I am asked to do a job without the requisite training.
2. I will sign-off task/job cards while in close proximity to the maintenance site so that I can physically verify all maintenance actions prior to signing them off.
3. If I find a discrepancy among two or more maintenance instructions, I will either seek a third party validation or I will use the most conservative option.

Tracking Personal Progress

Personal progress is all about moving from awareness to behavioral change. In our previous volume (Patankar & Taylor, 2004), we discussed four levels of risk:

1. *Good Samaritan Risk*: Risk is introduced in the system by the very act of maintenance. The more maintenance one performs, the higher the Good Samaritan Risk because every time a system is disassembled, there is a chance that it will not be re-assembled without an error.

2. *Normalized Risk*: Over a period of time, work-arounds are developed to compensate for ineffective maintenance procedures and every time such a work-around does not result in a negative consequence, a subtle reinforcement takes place in the mechanic's mind: perhaps the deviation from published procedure was okay.

3. *Stymie Risk*: Often, maintenance tasks involve disassembly or disconnection of parts/sub-systems in order to reach the specific part/subsystem that is called-out in the maintenance instructions. Every time a mechanic disturbs the configuration of an interfering part or system, he risks not returning that part or system to normal configuration. This risk is heightened by the fact that maintenance instructions do not specifically call for re-installation of all the parts that were removed/disconnected in the process of performing the target maintenance task; the instructions simply say 'return aircraft to normal'.

4. *Blatant Risk*: Risks in this category typically involve reckless behavior. Sometimes, such behaviors are attributable to latent conditions such as lack of appropriate training or equipment, poor norms throughout the organization, or inordinate performance pressures.

In order to manage the above risks, we have discussed pre- and post-task strategies in this book that would provide specific guidance on behavioral changes that need to take place. One way to measure the status of such changes, as well as to remind yourself about these issues, is to use pre- and post-task scorecards presented by us in our previous volume (Patankar & Taylor, 2004 Chapter 6). These scorecards require you to rate, on a scale of 1-5 (1 = Strongly Disagree, 2 = Disagree, 3 = Neutral, 4 = Agree, and 5 = Strongly Agree), the level of agreement with the following items:

- I am prepared to do this task because I have adequate knowledge, technical data, and tools/equipment to perform this task.
- I am mentally prepared to do this task (my stress, distractions, and time pressures are at a safe level).
- I am physically prepared to do this task (any physical discomforts such as aches and pains, fatigue, etc. are at a safe level).
- I have taken proper precautions to handle the work-management issues associated with this job: exercise tool control, make sure all the panels and access doors are securely fastened, and verify parts conformity with the IPC.
- I have practiced closed-loop communication to receive as well as give turnover about this task or to seek clarification about this task.
- I was assertive in identifying systemic deficiencies or errors as well as in acting on the deficiencies that were within my span of control.
- I practiced good teamwork because I was able to seek or give assistance to others, detect and rectify someone else's error, or resolve information/interpersonal conflict without increasing risk.
- I did not compromise my professional integrity because I documented all the work that I did and I did not sign for any work that I did not do.
- I took appropriate safety precautions to prevent personal injury to myself as well as others.
- I complied with all applicable regulations.

If such scorecards are used consistently, whether at an individual level or organizational level, they will serve as reminders as well as measurement tools to track individual/organizational progress.

Chapter Summary

In this chapter, we discussed assertiveness, teamwork, and professional integrity. Under assertiveness, our emphasis was on action-oriented assertiveness wherein we encourage you to actively listen to your colleagues as well as your own inner voice and act on any discrepancies that you may discover. Under teamwork, the emphasis was not only on helping each other improve the collective error-free performance, but also on sharing safety goals and building teams that encourage every member to practice the highest standard of professionalism. Under professional integrity, our emphasis was on maintaining consistency between words and actions and taking responsibility for any lapses in judgment or action.

Finally, we encourage the use of pre- and post-task cards to measure individual as well as organizational progress in developing safer work practices.

Review Questions

1. Describe one specific situation in which you practiced assertiveness by either active listening, using external third-party source to validate technical information, or speaking-up about safety hazards in your organization.
2. Discuss your perspective on teamwork in aviation maintenance by giving an example of one positive experience and one negative experience. Discuss the reasons why one experience was positive and one was negative.
3. Discuss some of the ways in which integrity could be improved at the organizational level.

Chapter 5

Professionalism and Trust: The Dual Mandate for Improvements in Safety

Instructional Objectives

Upon completion of this chapter, you should be able to accomplish the following:

1. Identify elements of professionalism in the aviation maintenance environment.
2. Describe the role of professional behavior in influencing the safety culture within a maintenance organization.
3. Discuss the paradox presented by the Speed-Accuracy-Tradeoff.
4. Identify elements of trust and distrust in the aviation maintenance environment.
5. Discuss the reasons for low trust levels between mechanics and their supervisors; propose ways to improve such trust.

Introduction

Crafts, trades and professions all form communities of practice that support creation and diffusion of tools and knowledge. These communities of practice also form subcultures where certain mutual expectations and values emerge. The concept of professionalism is distinguishable from those of craft or trade by the existence of a moral and ethical position to uphold, as well as a body of formal knowledge to be acquired—usually in a formal institution of higher learning. A physician, lawyer, teacher or aircraft mechanic differ in those ways from tailors, midwives, stonecutters, or IT programmers.

Professionals are distinguished from usual occupations by having a high degree of (a) competence in their field, (b) control (both in authority and ability) to make decisions based on their competence, (c) commitment to a greater public good, and (d) being central to the successful operation of the larger enterprise (Taylor & Felten, 1993, pp.132-136).

Most AMTs/AMEs around the world are employed by commercial airline carriers or by approved maintenance organizations/repair stations. Most of these people are professionals. They have attended many hours of school, stood rigorous examinations, and are thereafter certified by their country's government to approve as airworthy the aircraft they repair. Most maintenance management in the commercial aviation establishments have been promoted from the ranks—they have experience practicing the profession and continue to hold a certificate to practice. But their newly instilled support for productivity puts them in potential conflict with their mechanics. If we are to address safety and risk management among mechanics we must address the related issues of professionalism and trust first.

Professionalism, Self-control, and Communication Skills

Production deadlines such as flight times or heavy maintenance visit completion date are endemic in commercial aviation. These deadlines can be seen by mechanics as unreasonable or unattainable given the 'real world' within which they need to be accomplished versus the 'paper world' within which they are planned. It has been known and discussed for many years that airline management policy emphasis on meeting schedules can gradually compromise safety (Bruggink, 1985). This conflict between productivity and safety has been termed the 'Speed-Accuracy Trade-Off' or 'SATO' (Drury & Gramopadhye, 1991). Often SATO is resolved by delaying the aircraft's departure, transferring the work to a subsequent stop on the route, or by lowering quality. Any one of these alternatives causes feeling of 'stress' (or more correctly distress), which can lead to physical strain. Having information about priorities, together with control over the means of production to meet priority deadlines is a crucial factor in stress management and reduction. Airline mechanics often see their managers holding both the means or control as well as the required information. Mechanics thus see themselves as largely powerless to control this source of stress.

The professional AMT has been defined as embodying the joint characteristics of competence, centrality, control, and commitment regarding safety of flight (Taylor & Christensen, 1998, pp. 83-84). AMT professionalism is manifest in the exercise of these characteristics, together with a willingness to take responsibility for one's own behavior, to make judgments based on reliable data, and to assertively encourage this responsibility in others involved in flight safety.

Empowerment, employee involvement, and open communication have been shown effective in improving quality performance in North American industry (Taylor & Felten, 1993). Furthermore, the greater the focus on convergent purpose between the work unit and the enterprise, the greater will be the recognition of AMT's central responsibility and a positive effect of open communication. This lesson has begun to be applied in the management of aviation maintenance. Initial progress has already been made, but the lessons from those who are most successful are not widely known (Patankar & Taylor, 1999a).

Two factors—assertiveness and stress management—have emerged in our research as very effective in supporting professionalism among aviation mechanics. We have seen the control and the commitment of professional AMTs reinforced by their ability and willingness to be assertive—to speak up, even in the face of adversity and social pressure and to be able to listen to others under similar circumstances (Patankar & Taylor, 1999a). Professional aviation mechanics' willingness to actively manage and mitigate the effects of stresses and strains on their decision-making has also been seen to support competence and control to enhance their professionalism and their safety performance (Taylor & Christensen, 1998, pp. 135, 157).

Trust, Individualism And The Mechanics' Culture

We have found low trust levels of AMTs for their supervisors ranging from one-third to less than 10%, in a sample of maintenance organizations (Patankar, Taylor, & Goglia, 2002). This demonstrates that there is substantial variation among maintenance companies and that the problem of trusting one's foreman to support mechanic's concern for safety is a real one. And trusting one's supervisor is not the only trust problem. Our previous research has also shown that aircraft mechanics are more apt to be loners than other vocations (Taylor, 1999); and in the U.S. they tend to place less value on teamwork and MRM training than their counterparts in other countries (Taylor & Patankar, 1999a). This suggests that trust among coworkers, especially in the U.S. is lower than expected of professionals, and indeed lower value of coworkers' trust has been documented (Taylor & Thomas, 2003b). The implications of this problem are multi-faceted and long-term. For one thing, the incident and error investigation programs such as ASAP (FAA, 1997, 2002b) and MEDA (Allen & Rankin, 1995) require threshold levels of mutual trust among mechanics, and between management and AMTs that may not be met by many carriers. For another thing, because too many MRM training programs suffer from a lack of visible management support and encouragement (Taylor, 1998, 2000b; Taylor & Patankar, 2001), we believe that their long-term acceptance also requires mutual trust. Thirdly, only increasing this mutual trust will offer an avenue to resolving the industry-wide paradox formed by the apparent conflict between productivity and safety—otherwise termed the 'speed-accuracy trade off' or 'SATO' (Drury & Gramopadhye, 1991).

The Measures of Professionalism in Aviation Maintenance

Survey scales, measuring the value of assertiveness and the value of managing life's stressors, have been successfully developed and tested with mechanics over the past decade (Taylor, 2000a, 2000b). These two maintenance scales were derived from a larger set of measures originally created for evaluation of Cockpit Resource Management training for flight operations (Gregorich, Helmreich, & Wilhelm, 1990). The results from the stress management and assertiveness scales are of most interest to us here.

The two scales have been in use since 1991 and over 12,000 maintenance employees have completed them. They have been recently tested again for reliability and validity (Taylor & Thomas, 2003b). Our survey instrument, the Maintenance Resource Management/Technical Operations Questionnaire (MRM/TOQ), has been described elsewhere (Taylor, 2000a; Taylor & Patankar, 2001; Taylor & Thomas, 2003a, 2003b). It is also publicly available on the Internet at http://mrm.engr.scu.edu/ (successfully accessed on 3/18/04).

The data presented in the remainder of the chapter represent the most recent aircraft maintenance companies to participate in the MRM/TOQ survey during 2000-2003. Six samples are included. They are as follows:

- Company A (n = 116) is a 10% stratified random sample of the maintenance department of a large U.S. based airline.
- Company B (n = 129) is a group of volunteers from one maintenance department, prior to attending a human factors training program. The sample contains a large proportion of college-educated and female respondents, and is heavily weighted toward management respondents.
- Company C (n = 2,408) contains participants in a mandatory maintenance MRM training program in another large U.S. airline.
- Company D (n = 76) is all the maintenance employees in a smaller regional airline.

- Company E (n = 209) is from a large U.S. based aircraft repair station. It contains all maintenance management and a 10% random sample of mechanics.
- Company F (n = 1,035) includes participants in a mandatory maintenance MRM training program, in yet another large U.S. airline.

'Value of Recognizing Stress Effects'

This 'stress effects' attitude scale emphasizes the consideration of stressors at work and the possibility of compensating for them. Though not directly related to the theme of human communication or interpersonal relations, this factor proves to be an important concept for professional behavior of maintenance personnel. Mechanics, and many managers, coping with SATO find themselves frustrated and stressed. Recognition of that stress can lead to physical strain and poor decision-making, and it can create a powerful incentive to seek out and learn ways of coping with frustration and stress. This scale has been adapted from earlier work in evaluating aviation human factors programs (Taylor, 2000b; Taylor & Patankar, 2001). A high score means a high value placed on the consideration of stressors at work and the utility of compensating for them. Understanding and managing stress are typically individual coping activities. Improvement in awareness of stress management has been established in previous studies in terms of the marked increase in appreciation of stress management after training.

Figure 5.1 shows initial or baseline levels for importance of stress effects in the six companies recently surveyed. Post training and six-month follow-up surveys were also administered in companies 'C' and 'F', so those results are also shown.

The baseline mean scores for the six companies are all in the 'normal' range (45th to 55th percentile rank) compared with results collected 1991-2000. The lower mean scores indicate that respondents felt that stress and strains did not have as much effect on their quality of their work or their decisions. The two companies C and F both had higher baseline means than did the other companies, meaning they were more aware of stress effects than the others.

Companies C and F also received training, which included stress awareness and coping techniques. The stress management attitudes scores following training are also shown in Figure 5.1. The means for both companies increased significantly (Co.C, $F = 134.41$, $p < .000$; Co.F, $F = 136.99$, $p < .00$) immediately following training. Six months later the stress management attitudes are still significantly higher than the baseline for both companies although dropping back some.

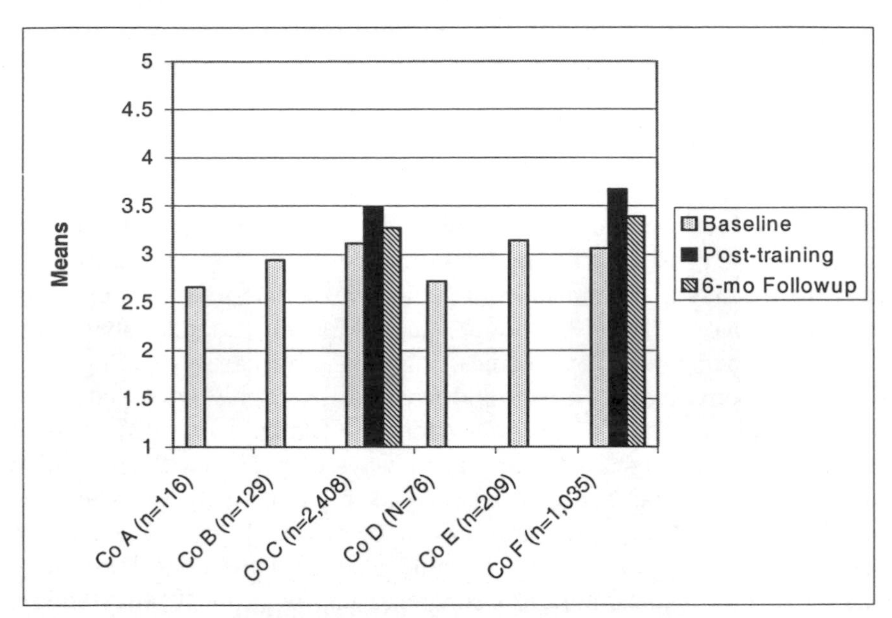

Figure 5.1: Importance of stress effects, by company

The messages from Figure 5.1 are several. First, it shows that there is some variation among companies in the amount of stress awareness there is (probably reflecting company culture as well). Second, there are immediate positive effects of training on maintenance employees' attitudes toward stress management, which appear to diminish only slightly over time. MRM training works in these cases and should be continued. To add strength to this argument, we know that these improvements in stress management attitudes and their associated behaviors have been strongly correlated with lower rates of injury and aircraft damage (Taylor, 1998; Taylor & Patankar, 2001).

There seems to be support for the benefit of life's experience in stress management as well.

With respect to purely individual differences—across all companies surveyed—we found a significant linear effect between respondent's age and 'Effect of my stress' (F = 2.74, df = 4, 2,905; p = .027) where this appreciation increased from the youngest to the oldest category.

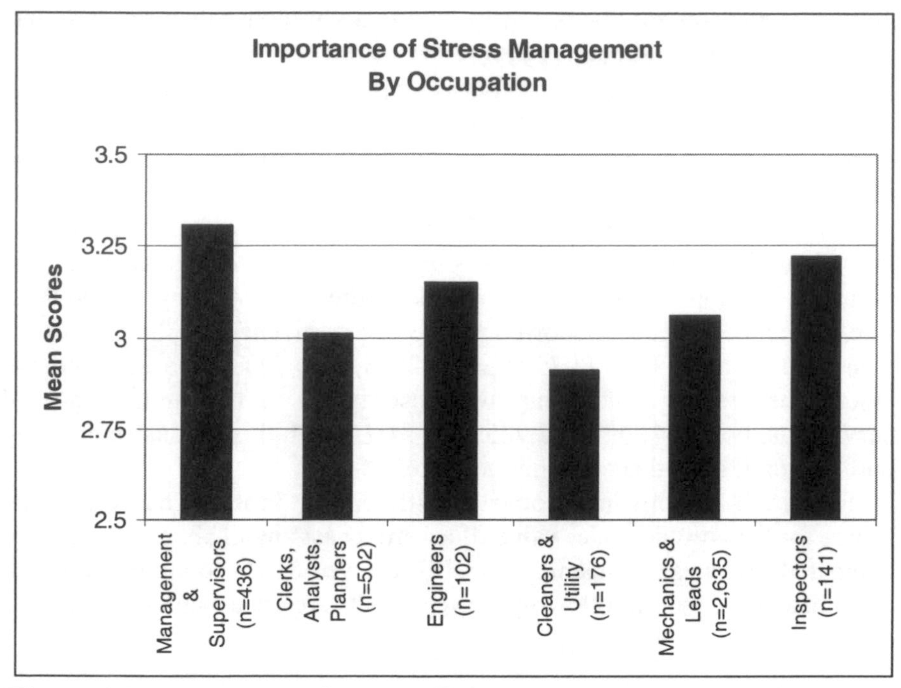

Figure 5.2: Importance of stress effects, by occupation

Figure 5.2 presents occupational differences in baseline attitudes toward stress management for six maintenance occupations across all six companies. Managers, quality inspectors and degreed engineers all score high on this scale. These groups tend to be older, higher paid and typically more thoroughly trained than the other three groups. Significantly lower scores are found for mechanics, clerical employees and aircraft cleaners ($F = 10.37$; $df = 5, 3983$; $p = .000$). Since mechanics represent such a large proportion of the employees of maintenance organizations these results suggest that MRM training will impact favorably on individual mechanics' stress awareness and coping abilities. In other words, this support of professional behavior can be effectively implemented through training.

'Value of Assertiveness'

This scale emphasizes the goal of candor and openness in maintenance communication. Assertiveness is a central concept to maintenance professionals and is an important part for human factors programs. A resultant high score on this scale emphasizes the value of speaking up and openness among maintenance personnel. Openness and honesty have proven to be important in maintenance human factors or MRM programs (Taylor, 1995; Taylor et al., 1997)

We found significant differences for assertiveness among the six occupational categories (F = 9.21; df = 5, 3976; p =.000). In particular the inspectors and mechanics value assertiveness significantly more than the other groups.

We also found significant effects of age and gender variables—individual differences that can be considered more independent of the industry. Main effects of age and gender were evident using MANOVA. The differences in gender showed 'value of assertiveness' to be greater for men than women; F = 7.07; df = 1, 2905; p = .008. The age and the assertiveness scale relationship was also found to be significant—and curvilinear, F = 3.51; df =4, 2,905; p = .007—with this attitude increasing with age until 45 and decreasing thereafter.

Figure 5.3 presents inter-company differences. In it, the baseline mean scores for the attitude scale 'value of assertiveness' in all six companies are compared. Because companies C and F received MRM training, their post training and six-month follow-up survey results are once again included.

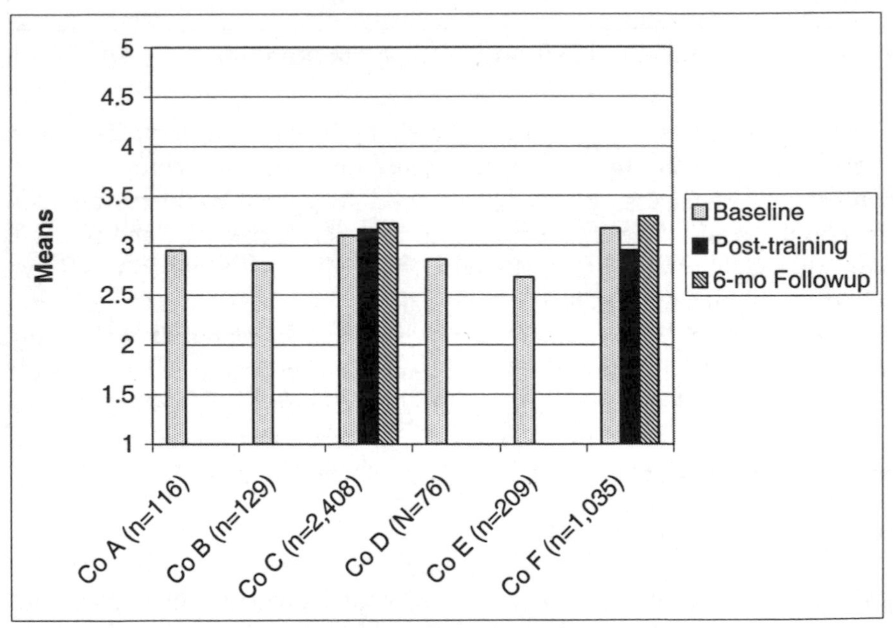

Figure 5.3: Value of assertiveness, by company

The baseline attitudes for these six companies span the lower to middle range of assertive attitude scores we have collected 1991-2000. For example, the baseline mean for company E is barely higher than the bottom third (i.e., the 33rd percentile) of all companies surveyed. That translates to a less than normal willingness to speak up if it might disturb or distress others. Companies C and F baseline scores are at the top of the 'normal' (middle third) range, while companies A, B, and D, are in the low-normal range.

There are interesting differences in the surveys collected after training in companies C and F. Company C's MRM training was different from that conducted in company F and those differences are reflected in the post-training results.

Company C's MRM training was divided into two phases, approximately six months apart. Both phases included communication topics. The first phase had a module on communicating more clearly. The second phase contained a module on assertiveness. As Figure 5.3 shows, Company C's assertiveness attitude increased significantly, immediately after training, then increased again six months later. This was consistent with our earliest findings that good communication skills training results in more positive attitude as well as increased communication (Taylor, 1995), about which we will say more below.

Company F's training, on the other hand, discussed the importance of good communication, but did not mention assertiveness nor provide any skills training to improve communication. In both regards, company F repeats the dominant pattern of MRM training in the U.S. during the 1990s (Taylor & Patankar, 2001). Most of that training left participants with the impression that 'good' communication meant 'getting along' and that 'speaking up', or assertive behavior might be seen as too abrasive or aggressive. The profile of company F's assertiveness attitude in Figure 5.3 —a drop in the value of assertiveness immediately after the training, followed by an increase in the months thereafter—is typical of those found elsewhere in the industry where little is done to deliver on the promise of open communication. The value of assertiveness, diminishing after training then increasing again six or twelve months later, has been noted in earlier studies and explained as the result of frustration with the apparent lack of progress in the MRM program in contrast with its obvious utility (Taylor, 1998; Taylor & Patankar, 2001).

Ironically, the professional attitude of valuing assertiveness has consistently shown positive relationships with subsequent safety outcomes when it has been encouraged and fostered (Taylor, 1995; Taylor, Robertson & Choi, 1997; Taylor & Patankar, 2001). It just has not been encouraged and supported very often in aviation maintenance.

An Example of Assertiveness Training Impacting Performance

Management sample: Early in the 1990s, one airline Vice President for Maintenance advocated assertive communication to his 1,000-plus subordinate managers and supervisors through their MRM training program. This training included a skill-building module in which participants role-played both the sender and receiver of assertive communication and they were coached to improve their abilities. The results of this innovative program have been documented (Taylor, 1995), as well as results several years later, after mechanics began to receive the same training (Taylor, Robertson & Choi, 1997).

For the managers, their average attitude on the assertiveness scale increased only slightly immediately after the training; but two months later their average showed a marked increase, which was sustained in two subsequent surveys six months and twelve months after training. This in itself is a good result, but it was in the correlation between this improved attitude and later performance that the program really succeeded. Performance indicators for on-time aircraft departure and lost-time injuries were collected monthly from the time of the training and for twenty-four months thereafter. Those two monthly performance measures over two years provided 48 data points to correlate with the managers' assertiveness attitudes for the post-training survey which preceded them. The two-month follow-up survey had two months less performance data after it, for a total of 44 data points combined over the two performance indicators. Likewise the six- and twelve-month surveys respectively had 36 and 24 performance data points following them. The assertiveness attitudes for the managers were correlated with their subsequent performance results of their own maintenance sections or work groups. The attitude scale and the two performance scales were coded such that a positive correlation meant that a manager's higher value of assertiveness was associated with more on-time departures or with lower frequency of lost-time injuries in his work group, and that lower attitudes were associated with fewer on-time departures and more injuries.

It was found that attitudes did indeed correlate positively and significantly ($p < .05$) with many of the possible points of subsequent performance. Just how those correlations were patterned is instructive—there are some spectacular spikes during those two years. One third (33%) of all 48 data points were significantly correlated with post-training assertiveness, which reflects a substantial effect of that attitude. For assertiveness two months after training, an incredible 80% of the 44 subsequent data points were significantly correlated. At six months the positive results disappeared to virtually zero (3%). Again at 12 months after training there were significant positive correlations with 63% of the 24 subsequent data points. This indicates that, at least for maintenance managers, the lessons about assertive communication learned from MRM training, coalesce and strengthen in the months afterward and stimulate behaviors that produce impressive performance effects. These results further show that the although positive effects of training on managers' attitudes appear to diminish over time, the specific effects of assertiveness—over 2/3 of all correlations are still significant twelve months afterwards—reveal how long-lived the effects of the training can be (Taylor, 1995).

AMT sample: After the MRM training had been completed for all managers, the trainers sought to provide the same training for AMTs. Common wisdom held that such communication training for mechanics and other hourly workers would be an unnecessary expense, especially in a period of prolonged financial hardship. It was felt by many in the company that the interpersonal communication training for managers had probably benefited the system as much as it could. At that time in the early 1990s, recurrent training for AMTs was limited to passively viewing videos produced by the company's Technical Training Department. But the trainers convinced the Vice President that improving assertiveness could continue and strengthen with mechanics. During the next six months, this program trained nearly 300 AMTs before it was terminated as a cost-saving effort. Unfortunately for the program (and the AMTs), the results we found were not available and computed until the following year.

As with their managers, AMTs completed pre- and post-training surveys, which provided comparisons for assertiveness and stress management. Unlike their managers, AMTs were not surveyed 2-, 6-, and 12-months after training. The pattern and level of attitudes pre-training and post-training are remarkably similar between the AMTs and Maintenance Management. For both groups a statistically significant increase was found following training for 'Recognizing Stressor Effects'. Like management, no significant difference was found for the 'Assertiveness' attitude index for the AMTs following training. The level of pre- and post-training attitudes is very similar for both samples for assertiveness. For 'Recognize Stress Effects', the AMT result reveals a lower measured level of sensitivity to the effects of stress than for the managers—but despite this, the AMTs' training is followed by a significant improvement in that attitude. To correlate attitude changes with performance, individual AMT's attitude data were combined into averages for the units to which they belong. These average attitude scores were rank-ordered and correlated with the rank-ordered maintenance performance data (which are available only by work unit), averaged over the six-month period following the onset of the AMT MRM training. 'Recognition of stress effects', and 'Assertiveness', were strongly related to lower incidence of aircraft ground damage. These relationships imply that the more assertive the AMTs are, and the better they recognize and manage stress, the fewer incidents of ground damage their work units experienced. The second and third performance indicators, 'occupational injury' and 'on-time departure', were not significantly related to AMT's post-training attitudes about assertiveness or stress management.

In the main, these results obtained for technicians parallel those reported above for their maintenance managers. Although they failed to replicate some earlier management results, these data are even stronger in showing positive effects of collaboration and MRM training (Taylor, Robertson & Choi, 1997). In particular, the findings show that the AMTs had stronger relationships than the managers between post-training attitudes and performance in the six months immediately following training. This is evidence for the fact that because AMTs are the persons directly affecting performance, their attitudes and subsequent behavior should most immediately relate to that performance.

AMTs who participated in this program were motivated by the training, but they never saw the quantitative results and statistics resulting from their efforts. For AMTs, the only real result from the MRM training was their disappointment when it was terminated. In this company thereafter, MRM was cynically known as, 'the flavor of the month—last month's'.

Trust and Aviation Maintenance

The boost to professionalism from stress management and open communication has been proven effective, but they are not sufficient in themselves. We believe that interpersonal trust is also required for effective communication and risk management. Mutual trust among AMTs and other ground support personnel cannot be taken for granted and must be consciously supported and encouraged.

Many airlines are trying to improve their safety culture by emphasizing communication and professionalism, together with awareness of decision-making, employee participation, and effective safety systems. A low level of trust, which results in cynicism among AMTs that permanent results will be achieved, stymies many of these programs. To more fully understand the concept of safety culture, substantial research effort has been directed toward developing the concept and measurement of trust.

Interpersonal Trust as Concept and Measure

The concept: Researchers have confirmed that the concept of trust is bipolar (it includes 'distrust' and 'trust') and that trust is a generic concept that includes interpersonal trust as well as trust of technology (Jian, Bisantz & Drury, 1998). Furthermore, in understanding the dynamics of trust in organizations, one can variously focus on the macro-level or micro-level of theory and analysis (Kramer & Tyler, 1996).

From the macro-level, investigators answer questions about how trust is related to organizational dynamics or management. Examples of such questions are whether trust in an industry or company has declined or whether trust can be rebuilt.

The micro-level perspective of trust considers the psychology of the individual—why people trust, and what aspects most influence individual trust. From this micro-level, investigators posit that trust facilitates truthful communication, and leads to collaboration (Mishra, 1996). We are interested in this aspect to the degree variables like individual's age, and gender can influence trust.

The measures: Social science has many examples of questionnaire scales that measure micro-level trust as an attitude, or affective state (e.g., 'It is important that person is trustworthy'), or as an opinion or evaluation ('this person is trustworthy'). Those reported scales that are found to rate high in construct validity, and reliability are usually developed and tested using samples of undergraduate students. In use, they emphasize a belief of trustworthiness (the degree to which others are seen as moral, honest and reliable) (Wrightsman, 1974).

We have recently developed measures of trust specifically for aviation mechanics (Taylor & Thomas, 2003b). These measures have proved to have good psychometric characteristics and are simple to collect and analyze. In our development of trust measures for aviation maintenance personnel both attitude and opinion measures for trust were considered and two reliable and valid scales were ultimately the result. One is the opinion scale, 'My supervisor's safety practices are trustworthy', and one is the attitude scale, 'I value coworker trust and communication'. These two new scales have been tested with over 4,500 mechanics, maintenance managers and support personnel. The trust scales, together with the 'stress management' and 'assertiveness' attitude scales described above, are the basis of the newest version of the MRM/TOQ. The six aircraft maintenance companies surveyed in 2000-2003 and described above provide the data for the trust results described below.

At the macro level, one of these trust scales reveals substantial differences among the six companies which are attributed to differences in safety culture and practice. They also reveal differences among different maintenance jobs and hierarchical levels, which are attributed to differences in occupational culture. There are also some differences found for gender and age.

The survey data from the six companies were analyzed to determine the level and characteristics of mutual trust.

Level of Trust for Supervisor's Safety Practices

This scale is comprised of five items that suggested trust of one's supervisor with regard to ethical behavior and safety practices involving the superior-subordinate relationship. The items are 'My supervisor can be trusted', ' My safety ideas would be acted on if reported to my supervisor', 'My supervisor protects confidential information', 'I know proper channels to report safety issues', and 'Mechanics ideas go up the line' (Taylor & Thomas, 2003b). Endorsement of the five items combined into this scale implies a favorable opinion toward a superior's trustworthiness in support of safety.

Inter-company differences: Significant differences were found for 'supervisor trust and safety'. Figure 5.4 shows mean scores for that trust scale among the six companies. A statistically significant ANOVA 'F' score was found for the Supervisor Trust and Safety scale ($F = 44.41$, $p < .000$).

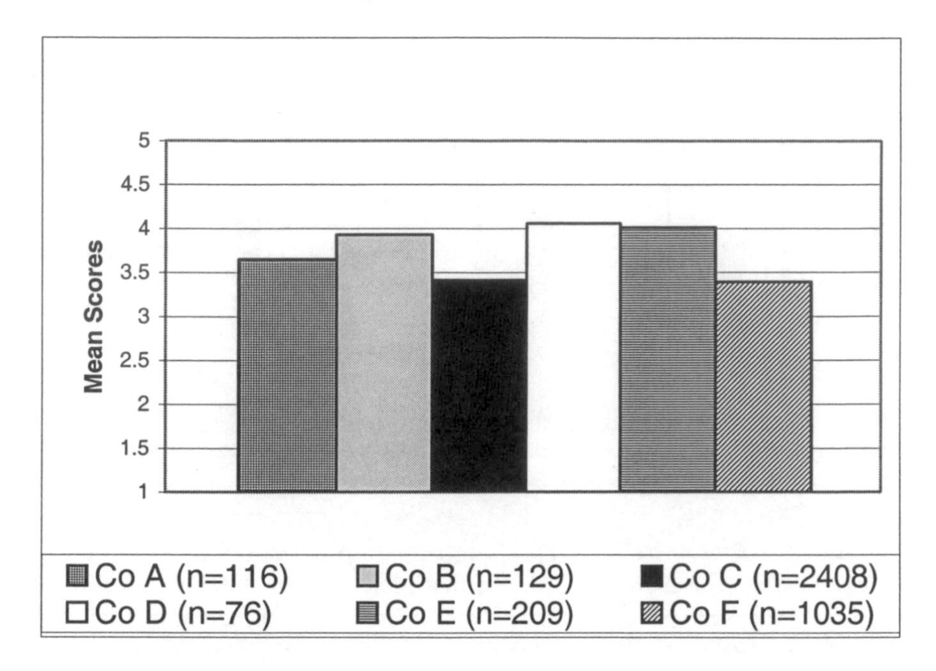

Figure 5.4: Responses to 'Trust Supervisor's Safety Practices', by company

Across the six companies, we find a considerable range of means, with a high score of 4.06 (Co. D) and low of 3.34 (Co. F) for the scale 'supervisor is trustworthy regarding safety issues'. When we examine only AMTs' mean responses, company D (a small regional carrier) is still highest (average = 3.95), and companies C and F (the largest airlines in our sample) show the lowest trust with averages = 3.26 and 3.21 respectively.

Expressed in terms of percent agreeing versus those disagreeing with the statements about supervisors' trustworthiness these differences among companies are dramatic (see Figure 5.5).

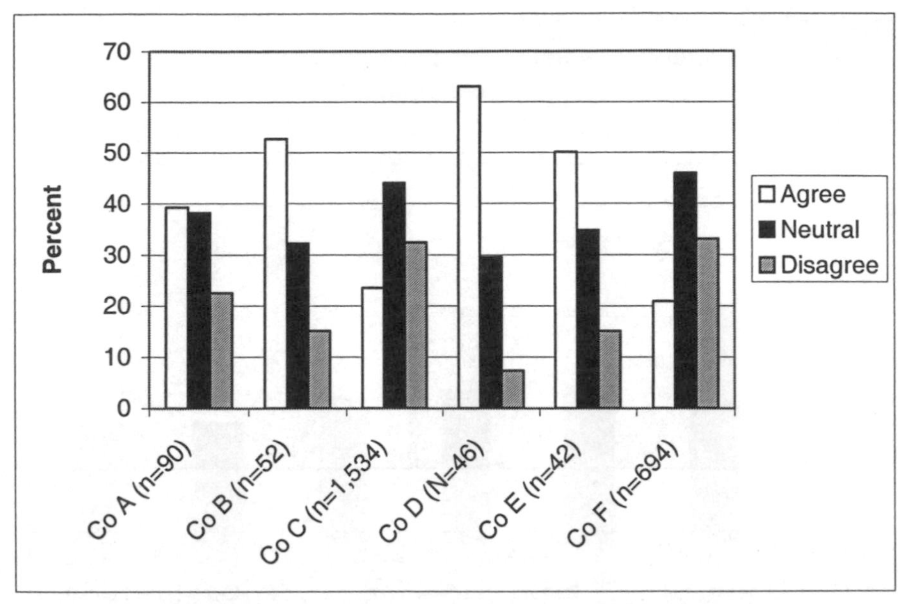

Figure 5.5: Responses to 'Trust Supervisor's Safety Practices', mechanic sample only

Mechanics and leads only, six companies: Figure 5.5 shows the percent of mechanics who answered 'completely agree', or 'agree', the percent of those answering 'neither agree nor disagree', and the percent saying 'disagree', or 'completely disagree'. From this we can see that across the six companies, a high of 63% and low of 24% mechanics say they are in agreement that their supervisor is trustworthy regarding safety issues—on the other hand, up to 31% of the mechanics in co 'C' say they disagree or strongly disagree with this. If we treat the 'neutral' responses as mechanics who are unable to endorse their supervisors' safety practices and combine them with those who 'disagree', then the total mechanics 'not trusting' in companies C and F are 76% and 74% respectively. This is clear evidence for differences in the various company safety cultures.

Occupational differences: In general, there is a sizeable perceived difference between mechanics and managers in their interpretation of their own supervisor's safety practices. Mechanics tend not to trust their managers as much as one might want in this high-risk industry. Figure 5.6 shows the mean scores among occupations for all six companies combined for the scale 'Trust Supervisor's Safety Practices'. The ANOVA 'F' score of 71.92 is significant $p < .000$).

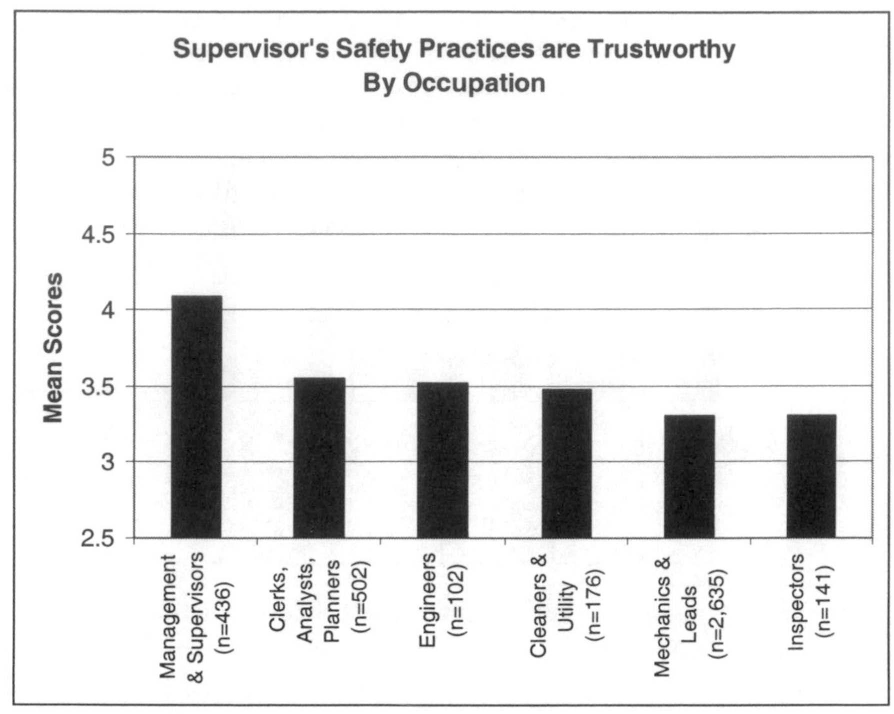

Figure 5.6: Responses to 'Trust Supervisor's Safety Practices' by occupation

Individual differences and interpersonal trust: Age and gender can be considered more independent of the industry variables and thus can be used to test the sensitivity of the two trust scales to individual differences. The supervisor trust scale showed a significant curvilinear effect for respondents of different ages (F = 4.13, df = 4, 2,905; p = .002)—an initially high level of trust for young respondents decreased with age until 45 years and then increased again. Both trust scales showed significant differences between men and women. These differences in gender showed women to have higher trust in their supervisor (F = 9.58, df =1, 2,905, p = .002), and also holding higher value of coworker trust (F = 4.86, df =1, 2,905, p = .028) as well.

Level of importance coworkers' trust and communication: This scale is comprised of five items that suggest a belief in trusting one's coworkers in their open communication in meetings and discussions. The specific questions were 'Having the trust of my coworkers is important', 'Debriefing after major task is important', 'Start of shift meetings are important', 'Others should make the effort for open communication', and

'Coworkers value consistency between words and action' (Taylor & Thomas, 2003b). Agreement with the five items related to this factor suggests a high value for mutual trust of coworkers in work-related discussions. Figure 5.7 presents and compares the baseline mean scores for the attitude scale 'value of coworker trust and communication' in all six companies. Because companies C and F received MRM training, their post-training and follow-up survey results are once again included.

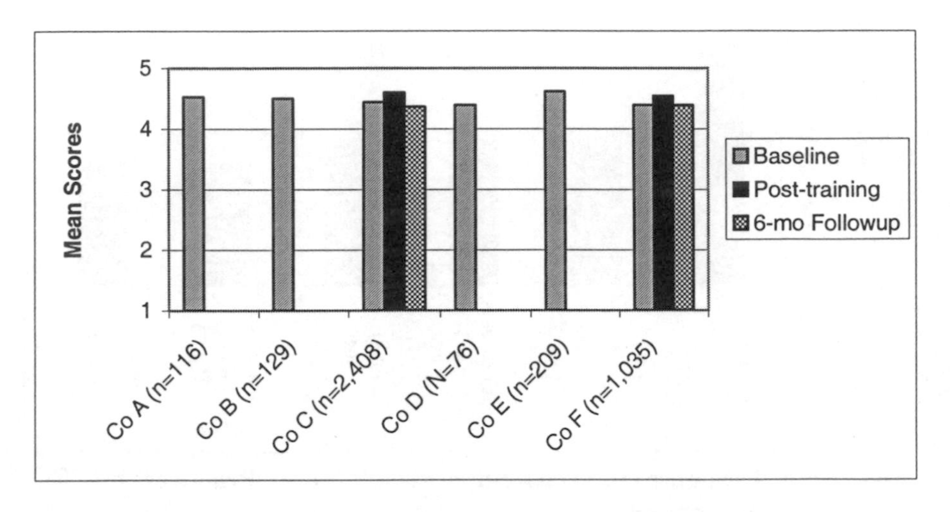

Figure 5.7: Value coworker trust and communication, by company

The inter-company comparisons among the six companies revealed no statistically significant differences among them. There were also no significant differences pre- and post-training for companies C and F. The coworker trust scale is stable and high—'it is indeed important to have mutual trust and communication with coworkers', is how this result from maintenance personnel in this group of companies can be interpreted.

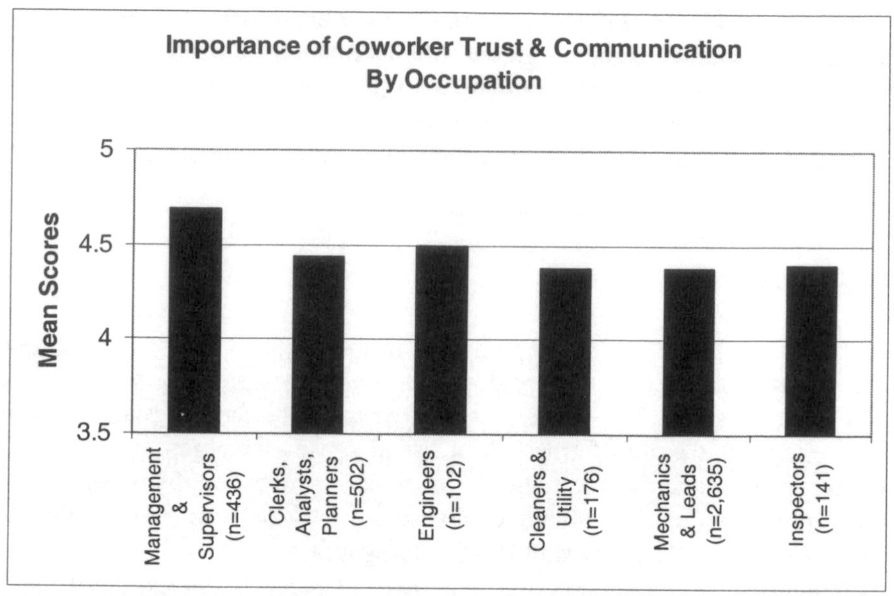

Figure 5.8: Value coworker trust and communication by occupation

We presented some trust scale data for five companies in our first volume (Patankar & Taylor, 2004 Chapter 5). Those data included the differences in 'value coworker trust' for six occupational categories. The present results, which include data from company F, further confirm those earlier findings. Figure 5.8 shows the mean scores for six occupational categories. Analysis of Variance revealed significant effects for occupational differences ($F = 20.54$; $df = 5, 3867$; $p = .000$). Managers had the highest mean score for this scale and mechanics, cleaners, and inspectors had the lowest mean scores.

Organizational differences: With these lower scores for AMTs as compared with managers, we tested differences in departments within the AMT group for all scales. Combining mechanics and inspectors into an AMT group, the main differences between Flight Line maintenance and Base Hangar maintenance departments across the six subject samples showed only one difference. Only this scale, 'value coworker trust and communication' revealed statistically significant difference ($p < .000$, $F = 20.1$; $df = 1, 1{,}521$). The other three scales are apparently not sensitive to the differences between the departments. Despite the fact that the Line Maintenance mean score for 'value of coworker trust and communication' is quite high (Mean = 4.33, Standard Deviation = .603, n = 783), it is still significantly below that of Base Maintenance (Mean = 4.47, Standard

Deviation= .544, n = 740). AMTs in the base hangars tend to be assigned to work together on complex jobs lasting as much as a week, while AMTs in flight line tend to be assigned to work by themselves on much shorter jobs. These conditions may well engender greatest value for collaboration among the base-hangar AMTs and the lesser value for this attribute on the flight line.

Discussion

Although the intent of MRM programs in the industry was to go beyond classroom instruction, such progress has not been achieved in a strategic and cohesive sense. Nonetheless, it is clear from the data presented in this chapter that the issue of trust—in a broad sense—emerges as a possible factor in influencing the degree to which professionalism and error mitigation programs are effective. Trust can only be won by practice and experience. It will never result from training alone.

Trust between mechanic and management groups was explored using survey methods. The scale 'Supervisor trust and safety' incorporates a trust of one's supervisor with regard to ethical behavior and safety practices. The scale 'Value coworker trust and communication' expresses a belief in trusting one's coworkers' communication in meetings and discussions. These two scales do support the assertion that aviation maintenance people find interpersonal trust to be a central concept in human factors.

In our earlier volume (Patankar & Taylor, 2004 p. 121), we reported that poor procedures, lack of training, and maintenance management's policies—factors that are outside an individual mechanic's span of control—account for as many as 16% of unairworthy dispatches. Despite exhortations by MRM trainers to 'speak up and improve open communication to change or to challenge poor procedure or policy', AMTs typically do not obtain skills from that training to do this. If they are waiting for their management to make the first move to 'open communication' they may be waiting in vain.

Poor procedures, lack of training, and management policy are organizational issues that may contribute toward reduction of mutual trust. For example, the authors have noted during their field observations that some companies have established a specific protocol for the communication of safety information in practice; however, the effectiveness of that protocol is unclear. Consequently, continued use of such safety programs is affected because the users do not receive a meaningful feedback. Over a period of time, such degradation of communication channels have led to lack of trust for the management's safety priorities. Errors will occur if mechanics use out-dated procedures because the incorrect procedures were not updated, even after the problem was voiced by mechanics. In this case again, AMTs are not likely to believe that their managers value quality and safety. Also, the familiar SATO paradox and the not uncommon practice of management signing for work completed when AMTs question doing so are sure ways of further reducing trust between mechanics and their managers. Therefore, it is not surprising that the MRM/TOQ survey data indicate that, for some companies, as few as a third (31%) of all maintenance employees trust their superiors—for mechanics alone, that trust sinks to less than a quarter (24%).

Conclusion

The professional AMT has several proven tools to help support his/her ability to manage risk. Both passive (e.g., stress management) and active (e.g., assertive communication) tools can be effective in minimizing the impact of current errors and minimizing the probability of future errors; however, the notion of trust plays an enabling role in the maintenance process. Without improving trust of management in the present aviation maintenance industry, professional mechanics have little chance to have more than periodic and superficial effect on error reduction. Mechanics must trust that their data are current and complete, the test or troubleshooting procedures are effective in detecting the defects that they are expected to detect, the spare parts that they use are of high quality and in appropriate configuration, and that their managers will act in the interest of safety. The managers must trust that the mechanics have, in fact, performed their tasks to the best of their ability when they sign-off an item on the task card. The pilots as well as the general public must trust that the aircraft that are delivered to them are, in fact, legally airworthy as well as technically sound.

Chapter Summary

Aviation maintenance professionals have tools and skills to develop and share, and ethics to uphold. In commercial aviation establishments many of these same professionals become supervisors and managers of others. Over time, the commercial considerations for customer service and profit present those managers with the SATO paradox—that in many cases becomes an ethical dilemma. Supervisors find themselves in support of productivity for company's survival while mechanics find themselves in support of safety for the same reason. A professional mechanic's trust in his/her supervisor's safety practices rests with experience and expectations. Either the SATO paradox is resolved to the satisfaction of both parties or a climate of mistrust and disbelief begins to form. Trust in, and support for, others' safety practices are core attitudes and behaviors in aviation maintenance.

Review Questions

1. Discuss why you consider yourself a 'professional'. Use the concepts of assertiveness, stress management, communication, and ethics discussed in this book to delineate your arguments.
2. Provide an example of 'trust'. Consider both interpersonal trust as well as trust in the aviation maintenance system. Use your own anecdotal examples to illustrate your points.
3. Discuss your interpretation of 'trust'. Consider both interpersonal aspects of trust as well as trust in the aviation maintenance system. Use your own anecdotal examples to illustrate your points.

Chapter 6

Safety Data Collection and Analysis

Instructional Objectives

Upon completing this chapter, you should be able to accomplish the following:

1. Describe the basic methods of analyzing the safety health of an organization.
2. Provide a brief overview of tools available to analyze qualitative and quantitative safety data.
3. Describe the human subjects concerns that need to be addressed in research projects that solicit survey data, error reports, and interviews.

Introduction

Due to the high reliability of the aviation system, it is clear that (a) accidents occur relatively infrequently, (b) most accidents are preceded by a number of reportable incidents with similar root causes and they in turn are preceded by even larger number of unreported incidents (the classic Heinrich ratio), (c) in order to minimize the *rate* of accidents in the future, the number of reported as well as unreported incidents need to be reduced, and (d) reliable matrices need to be developed to track problems, corresponding solutions, and the effectiveness of such solutions.

In this chapter, we will discuss the various types of data that are collected by maintenance organizations and how they may be used to reduce the number of accidents. We will also briefly discuss issues related to confidentiality of such data and the roles of various stake-holders in achieving systemic improvements.

What Types of Data Can You Collect?

The types of data that you collect will depend on the types of changes that you *want* to implement or you *can* implement, considering your resources and support from your management. The golden rule of data collection is as follows:

- *Do not collect data if you are not empowered to make the corresponding changes*

By collecting data without having the resources to implement corresponding corrective actions you are likely to lose the trust of your employees. Regardless of the size of your organization and the extant safety culture, the very fact that you are embarking on a data-collection mission sends a signal throughout the organization—something is about to change. So, if nothing changes, the employees are likely to lose trust in the organization's intent to change. Also, changes will have to be implemented on both individual-level as well as organization-level. Depending on prior experiences with change programs, the workforce might simply wait to see if the management changes or if there are any organization-level changes consistent with the data-collection effort prior to committing to changes in their personal work habits. In order to effect a comprehensive change that is aimed at minimizing future occurrence of similar maintenance errors, David Marx suggests that 80% of the changes should be at the systemic level and 20% of the changes should be at the individual level (Marx, 2000). With this advice in mind, and due consideration to the available budget, it might be most appropriate to start with a relatively small project wherein progress could be demonstrated at both organizational as well as individual levels.

Once it is understood that change is inevitable and resources need to be committed, it is time to consider the types of data to be collected. The data types can be classified into the following categories: performance data, event investigation data, and intervention program effectiveness data.

Performance Data

Performance data are all the data that help understand where the money goes—what are your losses? Are you losing money because of damage to aircraft during ground movement, are your employees getting hurt on the job, are your mechanics making errors in their write-ups to the extent that aircraft are being released in an unairworthy status, or are there any other events that affect safety of the equipment or personnel? Information about all such events will constitute performance data. Almost every aviation organization maintains at least the following types of performance data:

1. *Ground damage data*: Any time there is a ground damage incident wherein an aircraft is damaged by a ground vehicle or some other obstacle, significant expenses are incurred in returning the aircraft to service. Someone in the company keeps track of these costs. Unfortunately, the department or the persons who keep track of such costs may never speak with maintenance managers and consequently a comprehensive corrective action may be close to impossible.

2. *Lost-time injury*: When an employee suffers injury at work and is forced to take a certain number of days off (based on OSHA Requirement—29CFR §1904), that injury constitutes a lost-time injury because the worker is not productive during his absence. Such injuries therefore constitute a loss of production as well as cost of medical expenses. In some airlines, records of occupational injuries are turned over to insurance companies that provide Workmen's Compensation coverage. Often, these data are subsequently unavailable to maintenance management.

3. *Repeat repairs data*: Sometimes, repairs or maintenance actions need to be repeated because they were not performed to a satisfactory level at the preceding opportunity. A specific aircraft's maintenance logs will reveal at least some such items. However, a regular tracking system will be able to identify abnormal repeat repairs. Again, it is essential to track such repairs because it costs time, money, tools, and parts to repeat them. Also, there is a chance that due to the incorrect repair, the aircraft was rendered unairworthy for the flights undertaken between the commission and the discovery of that error.

4. *Logbook error data*: Many mechanics/engineers are taught to write the minimum in the maintenance logbooks so that they are somewhat protected from violation by the regulatory authorities. Well, the tendency to write minimum information in a maintenance logbook could also lead to errors because incomplete or even incorrect information gets communicated. Logbook errors may include incomplete information, failure to provide reference to the maintenance manual, wrong information, missing signature, and so forth. With the advent of computerized logbooks in many aviation companies, it may be even easier to analyze these data.

Error Investigation Data

Many companies are advocating the shift from blame culture. They may not be quite ready to embrace a just culture, but they are at least trying to distance themselves from the industry's blame culture of the past (c.f. Patankar & Baines, 2003). Typically, an error investigation is triggered by self-disclosure by the mechanic (ASAP-related), voluntary disclosure by the company (management discovered the error and reported it under the FAA's Voluntary Disclosure Program), regulatory violation cited by an FAA inspector (starting with a Letter of Investigation), or an incident/accident. Since all such investigations are initiated in response to an event (known consequence of an error), they are considered to be reactive. Of course, there are ways to capture or contain errors before they have any significant negative consequences, and there are ways to investigate or track such errors. We will talk about these *proactive* error-tracking systems in more detail below.

In aviation maintenance, the error investigation process tends to resemble a network rather than the single linear chain that is more common to a flight-related investigation. The network has the effect of the error being in the center, with threads of causal relations to that maintenance error radiating out in various directions. The resulting causal network may include a wide variety of entities—some of which may even be outside the airline. Nonetheless, the investigators are charged with identifying the primary and contributing factors leading up to the event. Of the several error-classification taxonomies available, the maintenance error investigators seem to prefer the MEDA taxonomy developed by the Boeing Company. Goodrich Aerospace, for example, has converted the MEDA taxonomy into a computer database that can be used to enter and track error data in accordance with the MEDA causal factors.

In spite of MEDA's popularity, it must be clarified that MEDA is an error classification tool, not an error investigation tool. It is not an investigation tool because it does not really help the investigators build causal networks associated with the events that they are investigating. For example, consider the error report presented in Example 6.1 below.

Example 6.1 ASRS Report Number 326178 (Edited for clarity)

> Aircraft departed on first flight after heavy maintenance visit for APU change and aft pressure bulkhead damage repairs. Aircraft would not pressurize after takeoff. Flight returned to gate and was taken to hangar. After troubleshooting, outflow valve was found blocked in open position. Aircraft was found to have departed with improperly cleared MEL for

outflow valve. When aircraft arrived for heavy maintenance visit, outflow valve was on MEL. When visit was complete, MEL appears to have been improperly cleared with outflow valve left open. MEL was properly cleared, pressurization system operated normally and aircraft returned to service. I feel company needs better control, tracking, and clearing procedures for MEL operations. My involvement in this incident was as inspector in heavy maintenance. This particular pressurization item was not a required inspection item per our general maintenance manual. If it had been, 'another set of eyes' would have been available, possibly preventing this from happening.

Supplemental information [typically a report filed by another individual associated with the same event—we may see two reports from two mechanics, or one from a mechanic and one from an inspector, or one form a mechanic and one from a pilot—all on the same event] from ARSR Report Number 326175: I was assigned to troubleshoot an aircraft which had returned to the field with a pressurization failure. In my briefing it was reported to me that this aircraft may have been released without certain procedures accomplished to allow it to pressurize. Also reported to me was other mechanics from day shift attempted to pressurize and were unable. Once at the aircraft, I reviewed the aircraft logbook and took note of an MEL procedure accomplished in order for the aircraft to be maintenance ferry flown for repairs and that no entry showed those procedures reversed. As previous mechanics had also looked at the outflow valve and found a piece of rubber hose in left-hand outflow valve as per MEL procedures, I also carefully exercised both outflow valves and under lower section of left-hand outflow valve noted small black block or hose which I removed. I noted no other discrepancies. I then started the APU and proceeded to pressurize the aircraft. On the multifunction display unit there were no warnings or indications of a problem. I pressurized the aircraft to 4 psi and maintained that pressure. As no other history had indicated problems prior to the MEL, I closed panels that were opened to inspect outflow valves and made appropriate logbook entries to release aircraft for service.

According to the MEDA classification, one would extract the following data:

- Maintenance was performed during a heavy maintenance visit—**Base Maintenance**
- The error resulted in a **Gate Return**
- There was a Gate Return because the crew was **unable to pressurize the aircraft**
- Pressurization was not possible because the outflow valve was blocked in the open position; it was also indicated on an MEL—**system not reactivated**

- The pressurization system was not reactivated because apparently nobody reviewed the maintenance log to discover the MEL status—**Information not used**
- No further information is available regarding the causes/events leading up to this error.

From the above data, only the words marked in bold would be visible to the analyst: Base maintenance—Gate Return—Unable to pressurize the aircraft—system not reactivated—Information not used. As you can see, the above data are helpful in understanding the factors that may have contributed to the error, but they are not helpful in establishing a causal link. In the above case, it is clear that the pressurization system did not work because it was not reactivated, but 'information not used' is a generic category which tends to hide the issue of MEL interpretation. Also, such tools are susceptible to a certain degree of investigator bias—the conclusions of different investigators investigating the same event are likely to be significantly different. In addition to, or in place of, MEDA, one might want to consider tools such as TapRoot or Causation Trainer to help develop a causal analysis of error-producing factors.

The TapRoot Model (www.taproot.com) incorporates a question-answer based system designed to identify human, technical, and systemic factors responsible for failures/malfunctions. This system has been commonly used in Chemical Process Plants in the United States and is now receiving some attention among airlines. It is a unique tool because 'it provides both inductive and deductive techniques for systemic investigation of the fixable root cause problems'. As such, one can use this tool to do both reactive investigations as well as proactive safety assessments of any given socio-technical system. Some of the additional tools available with the TapRoot system allow the analysts to even track the effectiveness of comprehensive solutions that are developed in response to a particular incident investigation.

The Causation Trainer Model (now offered under Outcome Investigator Toolset: www.outcome-eng.com) provides a very simple way to analyze cause-and-effect relationship between various events leading up to an incident/accident. Basically, the causal factors are classified into three categories: human error, rule violation, and mechanical failure. The analyst is expected to keep building the causal tree until all the causal links are explained. Again, David Marx, the creator of Causation Trainer, suggest that the comprehensive solutions should be such that 80% of the emphasis is on solving systemic problems and 20% on solving individual-level problems.

Irrespective of which causal analysis or error classification system you choose, it tends to provide pre-identified factors (such as hardware, people, management, maintenance instructions, etc.), one or more of which must be selected by the analyst. By the very nature of such 'forced' selection, it is likely that certain factors that do not necessarily fit in one of the pre-established categories may get ignored or get combined with other higher level factors. Also, the classification process itself is subject to the interpretation by the individual analyst.

In the section on *data mining*, we will present some alternate/emerging analytical tools that can be used to conduct taxonomy-independent causal analysis.

Intervention Program Effectiveness Data

Once the performance data have been collected and analyzed, appropriate intervention programs will have to be developed, implemented, and evaluated. Typical intervention programs include the generic Maintenance Human Factors/Maintenance Resource Management (MRM) training programs as well as specific programs to solve known problems with issues such as shift-turnovers, aircraft run-up & taxi, parts labeling and processing, surveillance of outsourced maintenance, logbook errors, MEL resolution, etc.

When generic MRM training is used to raise the level of awareness about safety issues or human performance issues among the employees, an attitude and culture questionnaire such as the MRM/TOQ (Taylor & Thomas, 2003b) can be used to measure the response at the baseline, pre-training, post-training, three months following training, six months following training, and one year following training. Such longitudinal measurement allows the analyst to track changes in the responses that can be correlated with the changes in performance measurements. For example, you will be able to determine whether there was any change in the ground damage incidents from the time you implemented a specific ramp-safety training program. Of course, one has to be sensitive to the possibility of some other factor (such as a coincidental citation by the FAA) external to the training program that may have contributed to the improvement in ramp safety. To account for such an effect, we suggest the use of a 'causal operator' in the calculation of the overall return on investment from any training program (Patankar & Taylor, 2004 Chapter 8).

Correlation of Data

Correlating various sources and types of data to derive meaningful conclusions is perhaps the most challenging task. If you have spent some time on the quality assurance side of maintenance, you are intimately familiar with volumes of audit reports that you have collected over the years from various line stations and vendors. How do you go about distilling these reports into meaningful and practical information? How do you know that you are correct or to what degree are you correct in your conclusions? Obviously, going through a detailed example of how something like this could be accomplished is beyond the scope of this book, but we would like to present you with a basic introduction to some of the tools that are available to you.

1. *Basic statistical tools*: The MRM/TOQ survey questionnaire is available for you to use (Available at http://mrm.engr.scu.edu/). You can administer this questionnaire to a randomly selected population of your mechanics, typically consisting of 10% of your maintenance workforce. The results from this round of survey will provide a baseline of attitude and opinion measures. Then, immediately prior to a training intervention, the same survey should be administered to the training class. The results of this survey will constitute the pre-training attitudes and opinions. Next, immediately after the training, the post-training version of the survey should be administered to measure the post-training attitudes and opinions. Then, 3-, 6-, and 12-months past the training, the post-training questionnaire should be re-administered. Collectively, the baseline through the 12-month questionnaire will provide a robust measure of changes in attitudes and opinions as a result of the specific training. If you are interested in going one step further, it would be necessary to actually interview a sample of the mechanics to determine how they are using their training. Such interventions and observations will provide even more reliable measures of 'effects of training'. For more illustrative details on this methodology, we suggest the following sources:

    a. Taylor, J.C. (1995). Effects of communication & participation in aviation maintenance. In *Proceedings of the Eighth International Symposium on Aviation Psychology*. Columbus, Ohio, The Ohio State University, pp. 472-477.

b. Taylor, J.C., & Christensen, T.D. (1998). *Aviation Maintenance Resource Management: Improving communication*, Warrendale, PA: SAE Press.

c. Patankar, M.S., & Taylor, J. C. (2004). *Risk management and error reduction in aviation maintenance*. Aldershot, U.K.: Ashgate Publishing Limited.

d. Thomas, R.L. & Taylor, J.C. *Evaluating MRM Programs: The Use of Company- and Department-Level Percentile Ranks in Evaluating MRM Programs*. June, 2003. Accessed at http://mrm.engr.scu.edu/newtool.html on 3/31/04.

2. *Analysis of events using MEDA, HFACS, or similar error classification tools*: These tools are most useful to analyze existing event investigation reports. For example, if you have a depository of all the event investigation reports in the quality assurance department or in the labor union's office, you may obtain permission to review these records and analyze them using a MEDA or an HFACS-ME form. For the latest version of the MEDA form, contact the Boeing Company at http://www.boeing.com/commercial/flighttechservices/ftssafety03.htm. A training manual for the HFACS-ME is available at http://hfskyway.faa.gov/HFACS/ME/StudentGuide.pdf.

   Also, depending on your individual circumstances and needs, you may want to modify the existing forms to add more categories/subcategories. After all, the tool should be useful to you! In one study (Patankar & Taylor, 2001a), we modified the standard MEDA form to incorporate Reason's General Failure Types. As a result, we were able to analyze ASRS reports to identify the various organizational and individual factors contributing toward maintenance errors. According to that study, the distribution was exactly 50-50! The conclusion was that about 50% of the times, the mechanics committed the errors due to individual factors such as lack of awareness or complacency; whereas, the remaining 50% of the times, the mechanics committed the errors due to organizational factors such as lack of training or lack of resources.

   You might note that earlier we pointed to David Marx's recommendation that error prevention strategies should focus 80% of their resources/efforts on systemic or organizational issues and 20% on individual issues. Connecting this recommendation to the Patankar and Taylor (2001a) finding noted above, we estimate that if 80% of the resources were focused on organizational factors, they will automatically have a significant influence on the individual factors.

3. *Analysis of event reports using a text analysis program*: Early on, text analysis programs were simply used to analyze short sentences filled-out by survey respondents. Now, text analysis programs are being used in 'concept analysis' wherein large volumes of text, in a variety of formats, can be analyzed by software programs such as LexiQuest Mine, Text Analyst, or QUORUM to filter out key concepts that are discussed in the text, their inter-relationships, and the correlational strength of those relationships. As a result of such analysis, it is now possible to simply provide one incident report (or any number of incident reports) to identify clusters of concepts and their inter-relations. The beauty of this system is that it is not dependent on the pre-identified categories, and it is possible for the analyst to drill down and actually view the context in which certain concepts appear. Furthermore, because the concepts identified by such a system are directly derived from the data, they are more likely to identify previously unknown or unimportant factors that may have become more important over time. With the Trend Analysis tool available in LexiQuest Mine, it is possible to actually track the relative importance of concepts (based on frequency of occurrence or severity of the consequences) over time. A detailed example of how LexiQuest Mine can be used to analyze event reports is presented in the next section.

Data Mining: LexiQuest Mine as an Event Report Analysis Tool

In order to fully understand safety challenges in aviation, it is perhaps even more important to study event investigation reports, self-reports of maintenance errors, etc. than to study the accident reports. Majority of such reports are available in the form of qualitative data or text. Hence, it is essential to understand the use of some of the text analysis tools. Without such tools, the analysis will have to be done manually using one of the pre-existing or custom-made error classification tools. As noted earlier, the modern text analysis tools such as LexiQuest Mine (see http://www.spss.com/spssbi/lexiquest/mine.htm for additional product information) provide some very exciting possibilities; however, consider the following notes of caution associated with the analysis and interpretation of the results prior to embarking on a text analysis project.

1. A text analysis system has very little, if any, domain (in this case, aviation) knowledge. Therefore, it is important for the analyst to be able to interpret the results accurately. Otherwise, it will be a 'garbage in, garbage out' situation. Interpretation of the text analysis results is perhaps the most difficult task.
2. Text data in event reports is often incomplete, loaded with domain-specific acronyms and jargon, and unstructured. As such, the analyst will have to spend substantial time cleaning-up the data and preparing it for analysis by the software.
3. Text analysis is just one tool. Its results should be compared with those from other means such as performance data, survey questionnaire data, and actual field interviews prior to developing an intervention strategy.

How Does it Work?

Different text analysis programs will work slightly differently and may provide you with a little variety in the possible output or results formats; however, the core principle is fairly simple. Basically, the system scans the text (in Microsoft Word® format, for example) and looks for common nouns, proper nouns, and in some cases verbs. Then, it provides results that may include one or more of the following:

1. *List of most common word pairs*: For example, if your dataset reveals that *maintenance* is the most common word, and *maintenance* occurs near *procedures* most often, then *maintenance* and *procedures* will be the top word pair.

2. *Clusters of concepts or terms*: Continuing with the above example, if *maintenance* is the most common term or concept, then all other terms or concepts that are related to this one could be mapped. Suppose the other concepts are *procedures, resources, training, time pressure,* etc. Then, not only can these concepts be mapped, but also each concept within this cluster can be assigned as the central concept, giving you different views of the clusters. The correlation factor between any two related terms in the cluster provides an indication of the relative significance of the relationships. Again, depending on the way the original text is structured, these clusters may or may not be useful. For example, suppose each report is a self-report from a mechanic/engineer rather than an investigation report compiled by a neutral party. Such report is highly subjective and therefore difficult, if not impossible, to compare with another (highly subjective) report of a completely

different situation. Nonetheless, if well-structured reports are available, the text analysis program can provide very fast and meaningful analysis.

3. *Trends of concepts or terms*: As more reports are added to the central database, the software can run subsequent analyses. With each such analysis, a slightly different cluster map can be generated. Therefore, it is possible to view, over time, the trend in changes in the relative significance of word-pair relationships. For example, in the first quarter, the word pair *maintenance-procedures* may be most significant; whereas, in the subsequent quarter, the word pair *maintenance-resources* may become more significant. The fact that there is a different word pair that is most significant may be predicated on the number of reports received in that period, but it may also be indicative of a rising problem area. The analyst will have to drill-down to the actual reports associated with these two word pairs to determine what has changed and why.

Example: Analysis of ASRS Maintenance Reports Received during 1996-1997 Using LexiQuest Mine

The Aviation Safety Reporting System first started encouraging reports from aircraft mechanics in 1996. About 116 reports from 'ground crew' were received in 1996 and about 110 reports were received in 1997. We cleaned-up the narrative sections of these reports to spell-out abbreviations such as 'acft' to be 'aircraft' and 'mech' to be 'mechanic'. Example 6.2 contains the original narration and Example 6.3 contains the narration after our editorial changes in the data clean-up process.

Example 6.2: Original narration of ASRS Report Number 324626

> acft arrived at gate with #1 tire flat. another mech and myself,along with lead mech,proceeded to gather up proper tools and pertinent parts to change both #1 and #2 main tire assemblies.We then proceeded to jack acft up,and removed #1 tire first.We then removed and replaced #2 tire assembly.We then began initial torquing of both #1 and #2 axle nuts.We found #2 tire would spin ok,with brakes off,but #1 tire would not spin.We then removed #1 tire assembly and determined #1 brake assembly was locked in 'brakes on'pos.We then proceeded to remove and replace #1 brake assembly.We reinstalled #1 tire assembly and resumed with initial torquing of both axle nuts,per maint manual specs.We then applied final torquing of both axle nuts per maint manual specs.We installed all retaining bolts to both axle nuts and safetied them. We installed both axle

covers, and safetied them. We filled out all paperwork and made proper entries in acft logbook.We clred acft of all used parts and tools and we departed gate area. The acft taxied out for tkof. On tkof,#1 tire and brake came off acft and came to rest on rwy. Acft returned to arpt with no injuries to pax on board, and no damage to acft. inspection showed all bolts and covers safetied and in place—both inner and outer bearings destroyed. Also,last several threads of axle nut damaged. Callback conversation with rptr Revealed the following info: a subsequent hearing (to tackle the sit)was undertaken with the faa, manufacturer, and company engineers. The committee determined that mechanically, everything was on and in place. The prob stemmed from the fact that the fk10 gear does not have a positive lock, even after the locking bolts are in. Rptr states that acft such as the dc9 and the b727 are different because they do have a positive lock, the bolt continues through the locking nut and the axle, at the same time. It was determined that the locking bolts were not in far enough, even though they were torqued as specified. Rptr states that in the future, he would rather take a chance at damaging 2 bolts than having a wheel come off during tkof. There have been approx 10 other related probs with this acft and since this incident, the company maint procs have been lengthened. Now,the number of rotations and the number of threads are counted. Rotations and threads are required to match, and a minimum measurement through the nut must be achieved by the use of a 'no go'gauge. Supplemental info from can [ASRS Report Number] 324623:on tkof, l lndg main gear lost outer wheel. Inflt, damage was assessed. Flc did a low fly-by twr for visual confirmation, then burned off fuel for lndg. Emer was Declared. Normal lndg was made. Acft sustained no other damage. Pax and crew are fine.

Example 6.3: Edited version of narration from ASRS Report Number 324626

Aircraft arrived at gate with number one tire flat. Another mechanic and myself, along with lead mechanic, proceeded to gather up proper tools and pertinent parts to change both number one and number two main tire assemblies. We then proceeded to jack aircraft up, and removed number one tire first. We then removed and replaced number two tire assembly. We then began initial torquing of both number one and number two axle nuts. We found number two tire would spin ok, with brakes off, but the number one tire would not spin. We then removed number one tire assembly and determined number one brake assembly was locked in 'brakes on' position. We then proceeded to remove and replace number one brake assembly. We reinstalled number one tire assembly and resumed with initial torquing of both axle nuts, per maintenance manual specifications. We then applied final torquing of both axle nuts per maintenance manual specifications. We installed all retaining bolts to both

> axle nuts and safety-wired them. We installed both axle covers, and safety-wired them. We filled out all paperwork and made proper entries in aircraft logbook. We cleared aircraft of all used parts and tools and we departed gate area. The aircraft taxied out for takeoff. On takeoff, number one tire and brake came off aircraft and came to rest on runway. Aircraft returned to airport with no injuries to passengers on board, and no damage to aircraft. Inspection showed all bolts and covers safety-wired and in place—both inner and outer bearings destroyed. Also, last several threads of axle nut damaged.
>
> A subsequent hearing (to understand the situation) was undertaken with the FAA, manufacturer, and company engineers. The committee determined that mechanically, everything was on and in place. The problem stemmed from the fact that the Fokker-100 gear does not have a positive lock, even after the locking bolts are in. Aircraft such as the DC-9 and the B-727 are different because they do have a positive lock, the bolt continues through the locking nut and the axle, at the same time. It was determined that the locking bolts were not in far enough, even though they were torqued as specified. In the future, he would rather take a chance at damaging two bolts than having a wheel come off during takeoff. There have been approximately ten other related problems with this aircraft and since this incident, the company maintenance process has been lengthened. Now, the number of rotations and the number of threads are counted. Rotations and threads are required to match, and a minimum measurement through the nut must be achieved by the use of a 'no-go' gauge.

For simplicity, let us illustrate with just the one report (as in Example 6.3 above) what LexiQuest Mine will do with such information. Once entered into the LexiQuest Mine database, it generated 31 concepts. The top five concepts, in the order of frequency of their occurrence, are as follows:

- Aircraft (n = 10)
- Axle Nut (n = 4)
- Tire Assembly (n = 4)
- Initial Torquing (n = 2)
- Positive Lock (n = 2)

The above results indicate that the ASRS report has something to do with an axle nut, tire assembly, initial torquing, and positive lock.

LexiQuest Mine also offers a knowledge map of concepts. Such a map, as illustrated in Figure 6.1, depicts other concepts that are linked with the most frequently used concept. In this case, *aircraft* is linked with *removed and replaced number, positive lock, threads of axle, proper entries, tire assembly, removed number, jack aircraft, Fokker-100 gear,* and *FAA.*

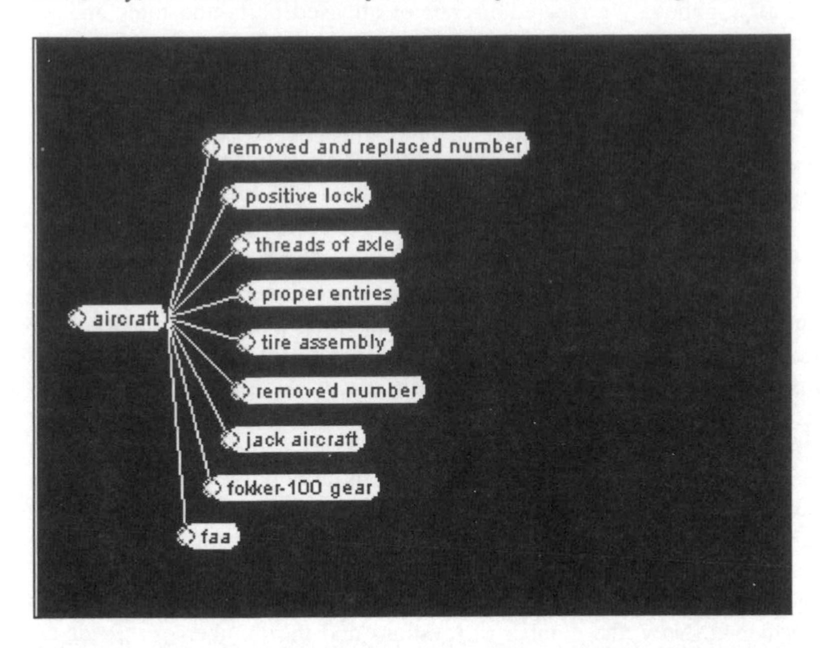

Figure 6.1: LexiQuest Mine's knowledge map of concepts derived from ASRS Report Number 324626

Notice that because the text was not structured enough to identify the precise problem, it would be almost impossible to identify the actual problem in this case. The list of concepts does not give any information to the analyst regarding the problem being reported. The analyst will have to drill-down deeper and actually read all the statements in which the significant concepts appear. Also, each concept such as *removed and replaced number, positive lock, threads of axle,* etc. could be easily reassigned to the 'entry point concept' position and all the associated links/concepts could be explored. Further, the signal strength could be adjusted so as to uncover deeper concepts—concepts that are not as frequently reported, but may be relevant/linked with more frequently reported concepts.

Method: To give you an example in this chapter, we analyzed ASRS maintenance reports from 1996 (n = 111) and 1997 (n = 93). The narratives from the original reports were cleaned-up, as discussed above. Knowledge maps for the combined dataset (n = 204) was analyzed to identify the most significant concepts and their inter-relationships. *Organize*, *Discover*, and *Track* algorithms were used to explore the relationships between the various concepts.

Results: Once the text files are loaded in the LexiQuest Mine system, it can conduct the data mining process and generate a list of concepts and relationships between them. LexiQuest Mine extracts unique noun-phrase concepts found in the document(s). The concepts are listed in a window from most frequent to least frequent (top 500 concepts are displayed). The top five concepts generated in this study are listed, but only the top concept is discussed using all three algorithms: *Organize*, *Discover*, and *Track*.

The top five concepts identified by LexiQuest Mine were as follows:

- Aircraft (203)
- Maintenance (125)
- Mechanics (24)
- FAA (21)
- Takeoff (18)

The frequency count of these concepts is indicated in the parenthesis. Only the top concept was analyzed using the knowledge map and its three available clustering algorithms. The results of those analyses were as follows:

1. *The Organize algorithm*: As we saw in Figure 6.1, the concept *aircraft* is shown to the left and concepts related to it are displayed via links. When we select the *Organize* algorithm, the concepts (all linked concepts) get clustered based on their proximity to the entry point concept—in our case, *aircraft.*

   For the purpose of this study, we selected a signal strength of 50 and up to 120 documents to be considered in the analysis. The greater the signal strength, the deeper the analysis—the number of related concepts increases and weaker concepts get displayed farther out.

   As you can see in Figure 6.2 below, there are four clusters of concepts related to the entry point concept *aircraft.*

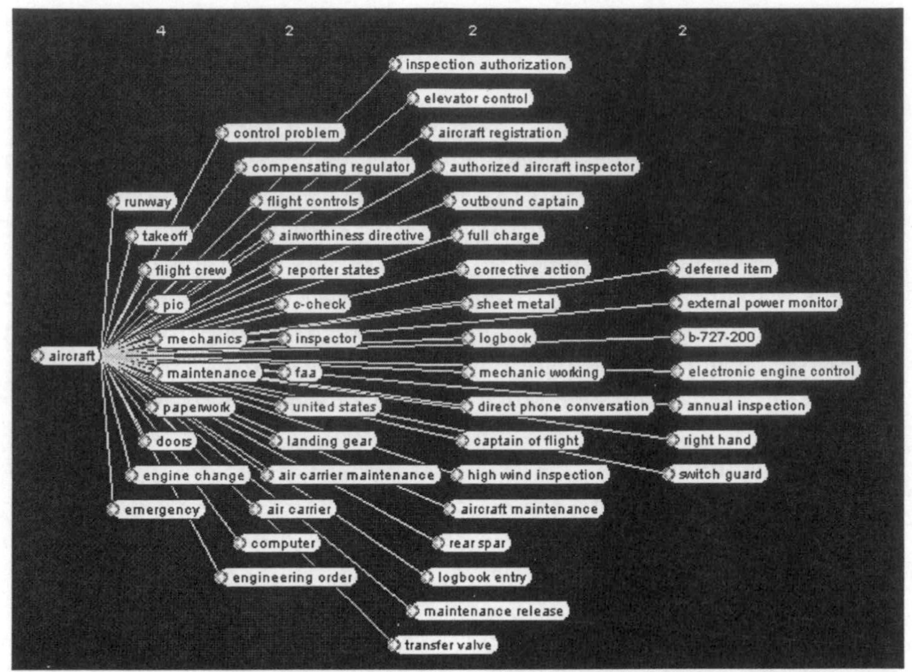

Figure 6.2: The Organize algorithm on 1996-1997 ASRS maintenance reports

At the top of each cluster, there is a number which specifies the strength of the weakest link between a concept in that cluster and the entry point concept. So, we can say that the weakest concept in the last cluster will have a strength of at least '2'.

The Organize algorithm is used to explore established semantic networks. That means, if we already know that the very fact that the concepts *aircraft* and *engine change* occur in close proximity is more significant than *aircraft* and *C-check* occurring at lesser proximity, then, the Organize algorithm would be helpful. It follows that since we are using voluntarily submitted error reports, we can say that errors related to engine change procedures tend to be reported more frequently than those related to C-checks.

2. *The Discover algorithm*: The Discover algorithm is more appropriate for exploratory research where precise relationships within word-pairs or significance of such relationships is not known. This algorithm allows the analyst/researcher to discover isolated, but linked, information among large volumes of data. This algorithm reveals exclusive relationships between two concepts.

Figure 6.3 shows the knowledge map under the Discover algorithm.

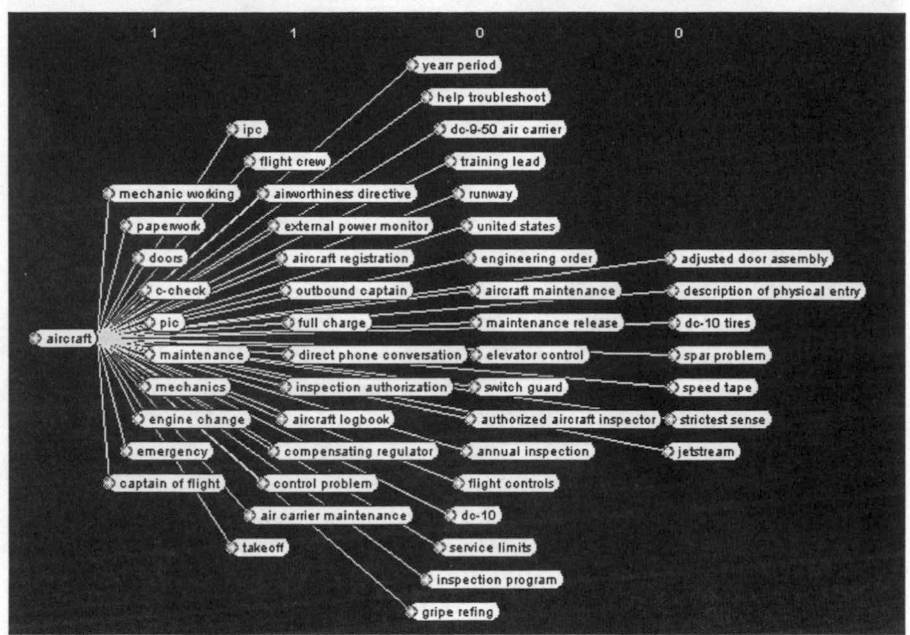

Figure 6.3: The Discover algorithm on 1996-1997 ASRS maintenance reports

As you can see now, the *C-check* and *engine change* concepts are both at the same proximity level to *aircraft*. Thus, one can conclude that the relationship between *C-check* and *aircraft* may be stronger than indicated previously by the Organize algorithm.

3. *The Track algorithm*: This algorithm is mostly useful in tracking elusive relationships among concepts. For example, as indicated in Figure 6.4, the concepts *aircraft* and *FAA inspector* seem to be closely related under the Track algorithm. These two concepts may not appear together much, but are certainly related. In order to determine the nature of the relationship among these concepts, one would have to drill deeper into the individual reports. In order to do that, the analyst has to simply select the two concepts and click on the *search* icon. A new window with hyperlinks to all the documents containing those concepts is returned.

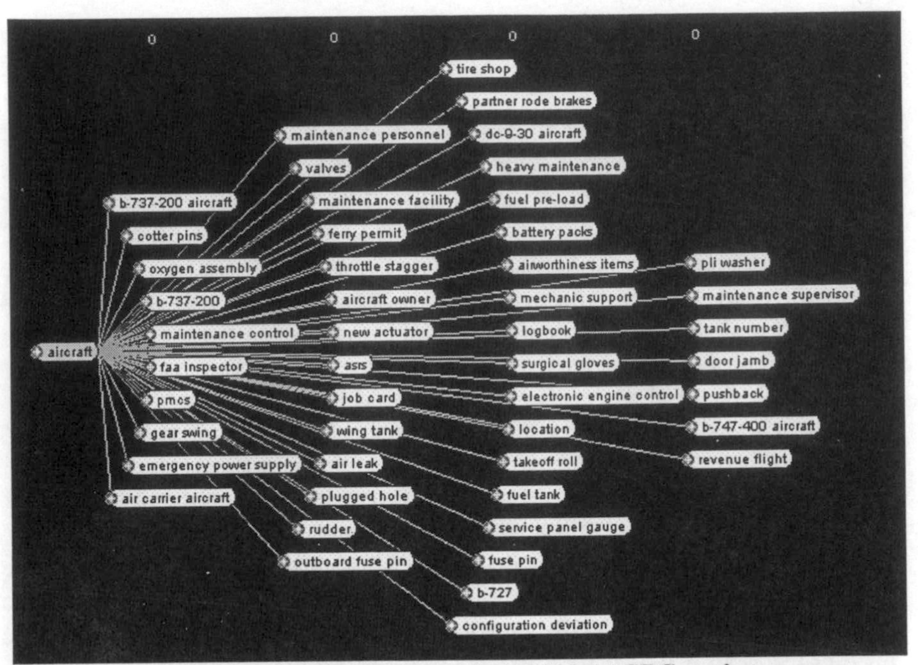

Figure 6.4: The Track algorithm on 1996-1997 ASRS maintenance reports

Discussion: Overall, a text analysis system such as the LexiQuest Mine system provides a very interesting and visual means to explore relationships between concepts embedded in large volumes of text files. Some of the key advantages of this system are as follows:

- The concepts generated by a text analysis system are driven by the source data and not pre-selected by the analysts. Hence, they are less susceptible to investigator bias and remain dynamically responsive to the changes in dataset.
- The trend analysis feature (not reviewed in this chapter) allows the analysts to understand the relative importance of certain concepts over time. In order to conduct such analysis, the LexiQuest Mine software date-stamps each report as it is added to the database. A concept that was not so important in the first year may become more important in the second year. Consequently, managers can prioritize their systemic improvement targets based on the most current and relevant data.

Ethics of Data Collection, Validity of Data, and Reliability of the Analyses

We can classify data collection efforts into anonymous, confidential, or public information. Data collected through survey instruments are typically anonymous, data collected through self-reporting programs are confidential, and data collected through documents such as FAA/NTSB accident reports are public information. As the aviation maintenance community strives to minimize maintenance errors through better data collection and analysis, it is imperative to understand the ethics of data collection, the validity of the data, and the reliability of the analyses.

Ethics of Data Collection

In academic institutions, any research activity involving humans falls under the purview of the Human Subjects Institutional Review Board (HS-IRB). Such a review board is entrusted with the responsibility to ensure that the research conducted by faculty members does not violate any of the basic human rights. Although the purpose of the HS-IRB and appropriate protocol to obtain IRB's approval are not new to the academic community, the practicing industry professionals may not be fully aware of these issues. Therefore, we would like to present the following key points.

Risks involved: Whenever a human is expected to participate in a research process, it is important to consider the various risks involved in both participating as well as not participating. For example, in soliciting responses to survey questionnaires what is the risk of participating? If there are no potentially incriminating or embarrassing items in the questionnaire and the sample size is large enough that the responses cannot be linked to specific individuals, the risk of participating may be minimal. Nonetheless, it is important to stress that the participation must be voluntary.

In the case of interviews, it is obvious that the person collecting the data knows who the subject is. The identity of this subject may not be a big secret, but the information provided by the subject must be protected. Such protection may be provided either through an explicit confidentiality agreement or by simply not recording any identifying information in the raw data and making sure that the sample size is sufficiently large.

What is the risk of not participating? It may be that the individuals who did not participate can somehow be identified. If so, is there any social pressure or risk to these individuals? Will their peers shun them because they did not participate in your study? These are legitimate questions and must be considered prior to data collection.

For confidential data, the most important risk is breach of confidentiality. What is the likely risk if the confidentiality is breached? Will the individual lose his/her job? Will the employer be under significant public scrutiny and incur financial losses? Such data have to be guarded very closely.

Typically, the basic risks are that the individual subject may lose his/her job or professional certificates, the organization may lose business or operating certificate, and the public may lose trust in the air transport system.

Data manipulation: If there is little trust between the employees and the management, data collection efforts are likely to be viewed with some suspicion. For example, in one company, the labor-management relations were so bad that the mechanics strongly believed that the management would alter raw data to hide the facts rather than accept responsibility and make changes. Also, these mechanics believed that the company's management was likely to provide corrupt data to the FAA and sacrifice the mechanics in an attempt to save the company.

One of the best ways to address such issues is to use an external party to collect and analyze the data. We suggest that the company have the external party execute a confidentiality agreement that is acceptable to labor as well as management. In some cases, we have noted another layer of confidentiality agreement that protects the identify of the individuals being investigated.

No change/feedback: Just as the risks of participating or not participating in a study are considered, the potential benefits to the subjects should also be considered. If no direct benefit or reward to the subjects/participants is anticipated, it should be clearly communicated. Too often, survey or interview data are collected and the individual subject has no way of finding out the results of his/her participation. If nothing changes and the subject believed that something was going to change or he was going to have some influence on the 'state of affairs', the subject is likely to be discouraged with the company's 'flavor of the month' improvement strategy.

In summary, make your intentions very clear, protect the raw data, de-identify reports and newsletter articles, and share the results/feedback candidly.

Validity of Data

With so much data available, how confident are you that your data are valid? For survey research, it is typical to ask four to five questions about the same general topic to gain a quantitative assessment of the level of agreement among the responses. But then, once in a while, you get those responses where someone decides to mark all 1s or all 5s without reading the question or worse yet, simply chooses the responses randomly. In both events, hopefully enough responses are collected so as to minimize the impact of rogue responses.

The more serious challenge comes in performance data and accident/incident investigation data. Beyond doubt, such data are subject to investigator bias. In the research community, typically multiple people are asked to investigate the problem independently (even if they are using the same recording instrument) and the variability in their interpretation/results is measured in terms of 'inter-rater reliability'. In the case of some internal event investigations, it is natural for management and labor parties to simultaneously investigate the event. If such investigations are a part of a collaborative effort such as the Aviation Safety Action Program, the final report will typically consist of a consensus rather than somewhat disparate individual reports. Throughout such event investigations, it is essential to seek *validation*, not *rationalization.* For example, when someone says that the torque is supposed to be 45 ft-lbs, it is essential to validate it as a fact with reference to an approved maintenance manual or equivalent, not take it as 'reasonable' because it seemed appropriate.

Reliability of the Analyses

As a safety program manager, you will face two basic questions:

- What are the most significant problems that deserve to be addressed within the resources that you have available?
- How can you test whether or not your intervention strategies are working?

General assessment of safety climate: In order to answer the above questions, you need reliable analyses of your data. In case you belong to an organization that has not had any significant negative experiences (regarding safety), you may want to start with a generic organizational safety culture survey (OSCQ, see Patankar, 2003b for the questionnaire) to determine the current status of attitudes and opinions regarding safety. For example, the OSCQ uses a scale called *Safety Opinions* (Patankar, 2003b). The following items were included in this category:

- 'Safety in this organization is largely due to good luck' (scored on reverse scale)
- 'Safety in this organization is largely due to adherence to the standard operating procedures'
- 'Safety in this organization is largely due to our collective commitment to safety'
- 'Safety in this organization is largely due to the efforts of a few key individuals' (scored on reverse scale)
- 'Safety in this organization is largely due to positive changes resulting from our past experience with incidents and/or accidents'
- 'Safety is my responsibility'.

Figure 6.5 shows the overall rating by the three employee groups at an airline on a scale of 1-5 (1 = strongly disagree, 2 = disagree, 3 = neutral, 4 = agree, and 5 = strongly agree).

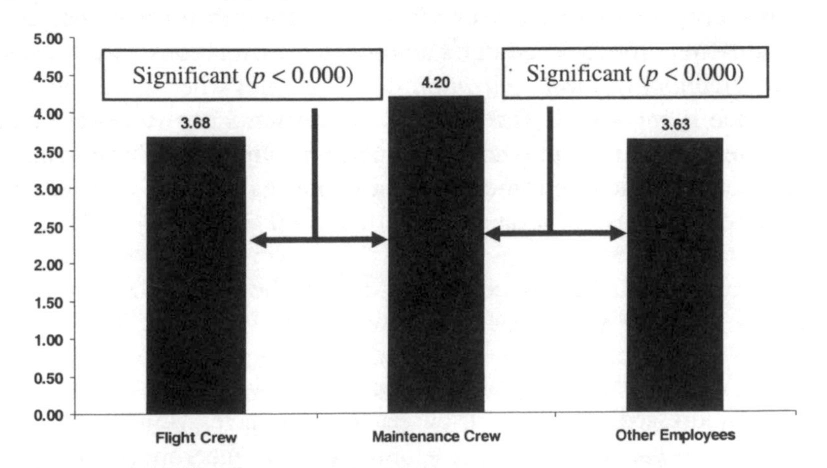

Figure 6.5: Comparison across the three employee groups on 'Safety Opinions'

Analysis of Variance (ANOVA) test on *Safety* revealed that the differences in the three employee groups' ratings were statistically significant ($p < 0.000$). Thus, there was a better than 99.99% probability that these ratings were not due to a random or chance error.

A further analysis of differences in ratings between groups, using the Scheffe test, revealed that there were statistically significant differences between flight crew and maintenance ($p < 0.000$) and between maintenance and other employees ($p < 0.000$). The maintenance crew rated the organizational safety at the highest level.

Similarly, other scales such as *Pride in Company* and *Supervisor Trust and Safety*, were also analyzed (c.f. Patankar, 2003b). Each scale was tested for reliability using *Cronbach's Alpha* (c.f. Gay & Airasian, 2003): Pride in Company tested at 0.88 (n = 397), Safety Opinions tested at 0.79 (n = 400), and Supervisor Trust and Safety tested at 0.85 (n = 389).

In a questionnaire such as the OSCQ or the MRM/TOQ, majority of the items need to be scored on a scale of 1 through 5. A statistical technique known as *factor analysis* allows the analysts to cluster the questionnaire items into groups such that responses to the items in that group are statistically similar. The level of similarity in the responses is used to determine the statistical reliability of the scale or the cluster. In the case of *Safety Opinions*, this reliability tested at .79: there is a 79% chance that the items in this cluster/scale are receiving similar responses. Since the general requirement for a scale to be reliable is Cronbach's Alpha of .70 or better, we can say that the Safety Opinion scale is reliable. Of course, the same questionnaire needs to be tested in multiple sample sets (in other organizations or other parts of the same organization) and the responses from different groups need to be compared for even more valid assessment of the reliability of the scale.

When surveys such as the OSCQ are repeated over a certain length of time (say 1-3 years) and over several different locations (base maintenance facilities, domestic line stations, overseas line stations etc.), the analyst will have a robust measure of the organization's *safety culture.*

Reactive, event-based measurement: Most of the time, safety managers are reacting to events such as accidents, regulatory violations, self-disclosures, lost-time injuries, etc. to determine what has gone wrong in their maintenance system and how it could be restored. In such circumstances, event investigation tools such as the Causation Trainer or the TapRoot System are very useful. These systems will help identify the individual-level as well as organization-level factors that contribute toward increase in risk in general or increase in specific events such as accidents, regulatory violations, and lost-time injuries.

As a result of several research studies in aviation maintenance over the past 15 years, it is now common knowledge that maintenance error tend to occur due to factors such as poor maintenance instructions, lack of resources, lack of training, distractions, schedule pressures, etc. So, you are not likely to find anything surprising in your organization; however, the next challenge to safety managers is about implementing effective comprehensive solutions to prevent similar errors for similar reasons. For example, suppose that in your organization, the biggest factor contributing to maintenance errors is poor maintenance instructions. You are also likely to discover that the process to get the maintenance instructions revised is cumbersome and ineffective. Consequently, the mechanics have developed several 'workarounds' and created norms. Then, the question is whether you have the resources to overhaul the document revision process in your organization. If the event that triggered this level of enquiry is an accident or regulatory violation or otherwise getting interest from your local regulatory authorities, the answer to the above question is likely to be positive; otherwise negative.

Introduced in 1998, Aviation Safety Action Partnership (ASAP) programs, have begun to be implemented in the United States. In such a program, employee organizations (e.g., labor unions), management, and FAA representatives come together to solve systemic problems that were discovered through sole source reports from company employees. Initial observations of ASAP programs at three aviation maintenance organizations indicate that the presence of FAA representatives as well as the high level of mutual trust that exists among the three parties provides strong encouragement for the senior company management to commit resources necessary to implement appropriate systemic changes. Moreover, as a part of the Memorandum of Understanding (MOU) required in order to initiate the ASAP program, there is an obligation to track the effectiveness of the systemic changes. (Patankar & Driscoll, 2004).

Correlation of performance data with intervention programs: The general expectation from senior management is that the intervention programs such as training, overhaul of the document revision process, or acquisition of new equipment should have a significant effect on corresponding performance criteria. For example, say your company averages about one logbook error per day that results in an unairworthy release. Such a release is likely to cost about $10,000 fine per flight until the error is corrected. So if you were to introduce a training program designed to minimize logbook errors, you will need to demonstrate that the logbook errors have diminished; and consequently, that you are also able to demonstrate cost savings.

With multiple intervention strategies—such as many different training programs, maintenance process changes, and management changes—it is difficult to attribute a specific value to the overall improvement in safety performance by the level of contribution made by a specific training program. Under such circumstances, we recommend the use of a 'causal operator' (Patankar & Taylor, 2004, Chapter 8) wherein due credit is given to multiple safety improvement efforts.

Chapter Summary

In this chapter, we have discussed a variety of safety assessment tools including survey questionnaires, performance data, and event investigation tools. Our emphasis has been on the introduction of how such tools could be used to assess systemic safety health in your organization so that you are able to develop the appropriate intervention strategies and measure the effectiveness of those strategies. Certain important ethical considerations are also discussed.

Review Questions

1. ASRS Maintenance Reports (ASRS) or Reports from the Confidential Human Error Investigation and Reporting Programme (CHIRP) provide an interesting perspective on the type of maintenance errors that occur in a normal working environment. Take a subset of such reports (say 10-20) and analyze them using MEDA, TapRoot, and/or Causation Trainer. Note your critique about each system.
2. If you have an internal, company-specific error-reporting program, you can repeat the analysis of Question 1 above using the internal error reports. Compare the errors discovered in Question 1 with those discovered in Question 2.
3. Collect your company's performance data such as lost-time injuries, ground damage statistics, flight delays over 15 minutes (due to maintenance), flight cancellations, inflight engine shut-downs, gate returns, and diversions. Obtain the corresponding cost values for each consequence and create a table of consequence versus cost per quarter/year.
4. Use a text analysis tool such as LexiQuest Mine or TextAnalyst to analyze error reports from the company's database or ASRS or CHIRP. A 30-day evaluation copy of the TextAnalyst software is available at http://www.megaputer.com.
5. Discuss the effectiveness of certain specific error-reduction programs in your company.

References

Allen, J.P., & Rankin, W.L. (1995). A summary of the use and impact of the Maintenance Error Decision Aid (MEDA) on the commercial aviation industry. In *Proceedings of the 48th Annual International Air Safety Seminar.* Seattle, WA: Flight Safety Foundation and International Federation of Airworthiness.

Ashby, W.R. (1956). *Introduction to cybernetics* (pp. 202-218). New York: John Wiley & Sons.

Badham, R., Clegg, C., & Wall, T. (1999). Sociotechnical theory. In W. Karwowski (Ed.), *International encyclopedia of ergonomics and human factors*. New York: Taylor & Francis.

Beabout, G., & Wennemann, D. (1994). *Applied professional ethics: A developmental approach for use with case studies*. Lanham, MD: University Press of America.

Boeing (2003, May). *Statistical summary of commercial jet airplane accidents worldwide operations 1959-2002*. Retrieved March 23, 2004 from http://www.boeing.com/news/techissues/pdf/statsum.pdf.

Boksanske, F. (2002, November 12). *A letter to the U.S. Department of Transportation Office of the Inspector General regarding Review of Air Carriers' Use of Repair Stations.* Retrieved March 20, 2004 from http://www.amfanatl.org/Pages/11_Safety&Standards/Ltr_DOT_11-08-02.html.

Bruggink, G.M. (1985). Uncovering the policy factor in accidents. *Air Line Pilot*, May, pp. 22-25.

CAA (2003, December). *CAP 716: Aviation maintenance human factors: Guidance material on the UK CAA interpretation of Part 145 Human Factors and Error Management requirements.* Retrieved on March 20, 2004 from http://www.caa.co.uk/docs/33/CAP716.PDF.

Chapanis, A. (1965). *Man-machine engineering*. Belmont, CA: Wadsworth.

CNN (2000, December 10). *Maintenance key focus on Alaska Airlines hearings.* Retrieved April 9, 2003 from http://www.cnn.com/2000/US/12/10/crash.alaska. reut/

Coakes, E., Willis, D., and Clarke, S. (Eds.) (2002). *Knowledge management in the sociotechnical world.* London: Springer-Verlag.

Dawson, D., & Reid, K. (1997). Fatigue, alcohol and performance impairment. *Nature*, vol.388, p. 235.

Drury, C. (1998). Work design. In M. Mattox (Ed.), *Human factors guide for aviation maintenance*. Retrieved August 20, 2002 from http://hfskyway.faa.gov/HFAMI.

Drury, C.G., & Gramopadhye, A. (1991). *Speed and Accuracy in Aircraft Inspection.* Position Paper for FAA Biomedical & Behavioral Sciences Division, Office of Aviation Medicine. Washington, D.C.

Eiff, G. (1999). *Human factors: safety and productivity equation, and return on investment*. Presented at the 1999 SAE Airframe/Engine Maintenance and Repair Conference. Vancouver, B.C.

FAA (1997). *Advisory Circular No. 120-66, Aviation Safety Action Partnerships (ASAP).* Washington, DC: Federal Aviation Administration.

FAA (2002). *Advisory Circular No. 120-66B, Aviation Safety Action Partnerships (ASAP).* Washington, DC: Federal Aviation Administration.

Gay, L.R., & Airasian, P.W. (2003). *Educational Research: Competencies for analysis and application* (7th ed.). Englewood Cliffs, NJ: Merrill, Prentice Hall.

Gilbreth, F.B. (1911), Motion study, New York: Van Nostrand.

Goldsby, R. (1996). Training and certification in the aircraft maintenance industry: Technician resources for the twenty-first century. In W. Sheperd (Ed.), *Human Factors in Aviation Maintenance—Phase Five Progress Report*. Washington, DC: Federal Aviation Administration/Office of Aviation Medicine.

Gregorich, S., Helmreich, R., & Wilhelm, J. (1990). The structure of cockpit management attitudes. *Journal of Applied Psychology*, 75, 682-690.

Hammer, M., & Champy, J. (1993). *Reengineering the corporation.* New York: Harper Business Press.

Hawkins, F.H. (1987). *Human factors in flight*. Aldershot, U.K.: Ashgate Publishing Limited.

Helmreich, R.L. (In Press). Culture, threat, and error: Assessing system safety. In *Safety in Aviation: The Management Commitment: Proceedings of a Conference*. London: Royal Aeronautical Society.

Helmreich, R., & Merritt, A. (1998). *Culture at work in aviation and medicine: National, organizational and professional influences*. Aldershot, U.K.: Ashgate Publishing Limited.

Helmreich, R.L., Klinect, J.R., & Wilhelm, J.A. (2001). System safety and threat and error management: The line operations safety audit (LOSA). In *Proceedings of the Eleventh International Symposium on Aviation Psychology*. Columbus, OH: The Ohio State University.

Jian, J., Bisantz, A.M., & Drury, C.G. (1998). Towards an empirically determined scale of trust in computerized systems: distinguishing concepts and types of trust. In *Proceedings of the Human Factors and Ergonomics Society 42nd Annual Meeting*. Santa Monica: HFES, 501-505.

Kanki, B.G., Marx, D., & Hale, M.J. (2004). Socio-technical Probabilistic Risk Assessment: Its capabilities and limitations. Accepted for inclusion in the *Proceedings of the International Conference on Probabilistic Safety Assessment & Management*. Berlin, June 14-18.

Kohlberg, L. (1984). *The psychology of moral development*. New York: Harper and Row.

Kramer, R.M., & Tyler, T.R. (1996). Whither Trust? In Kramer, R.M. & Tyler, T.R (Eds.), *Trust in organizations*. Thousand Oaks: Sage Publications.

Marx, D. (2000). *The causation trainer* (v.1.0) [Computer-based training program]. Retrieved August 28, 2001 from http://www.causationtrainer.com.

Marx, D.A., & Slonim, A.D. (2003). Assessing patient safety risk before the injury occurs: an introduction to sociotechnical probabilistic risk modeling in health care. *Qual Saf Health Care*, *12* (Suppl II), pp. 33-38.

MEDA (1994). *Maintenance Error Decision Aid*. [Results Form]. Collaborative effort of The Boeing Company, British Airways, Continental Airlines, United Airlines, the Federal Aviation Administration, and the International Association of Machinists.

Mishra, A.K. (1996). Organizational responses to crisis: The centrality of trust. In Kramer, R.M. & Tyler, T.R (Eds.), *Trust in organizations*. Thousand Oaks: Sage Publications.

Patankar, M.S. (1999). Professional and organizational barriers in implementing maintenance resource management programs: Differences between airlines in the United States and India. SAE Technical Paper 1999-01-2979. In *Proceedings of the SAE Airframe/Engine Maintenance and Repair Conference.* Vancouver, B.C.

Patankar, M. (2002). Causal-comparative analysis of self-reported and FAA rule violation datasets among aircraft mechanics. *International Journal of Applied Aviation Studies* 2(2), 87-100. Oklahoma City, OK: FAA Academy.

Patankar, M. (2003a). [An interview with Mr. David Marx on November 6, 2003]. Unpublished raw data.

Patankar, M. (2003b). A study of safety culture at an aviation organization. *International Journal of Applied Aviation Studies 3*(2), 243-258. Oklahoma City, OK: FAA Academy.

Patankar, M., & Baines, K. (2003). Establishing effective error-reporting programs: A cross-cultural comparison of lessons learned. In the *Proceedings of the 17th Annual FAA / CAA / Transport Canada Safety Management in Aviation Maintenance Symposium: Integrating Human Factors Principles.* Toronto, Canada. September 16-18.

Patankar, M., & Driscoll, D. (2004). Preliminary analysis of Aviation Safety Action Programs in aviation maintenance. In M. Patankar (Ed.) *Proceedings of the First Safety Across High-Consequence Industries Conference*, [CD-ROM]. St. Louis, MO, pp. 97-102.

Patankar, M., & Northam, G. (2003a). A study of student pilot attitudes and behaviors. In R. Jensen (Ed.) *Proceedings of the Twelfth International Symposium on Aviation Psychology*, [CD-ROM], at Ohio State University, Dayton, OH.

Patankar, M., & Northam, J. (2003b). Beyond compliance-based safety: A discussion of safety attitudes and behaviors among flight students. In the *Proceedings of the XVth Triennial Congress of the International Ergonomics Association.* Seoul, South Korea. August 24-29, 2003.

Patankar, M., & Taylor, J. (1999a). Professional and organizational barriers to implementing macro human factors based safety initiatives:

a comparison between the airlines of the United States and India. In *Proceedings of the SAE Airframe/Engine Maintenance and Repair Conference* [SAE Technical Paper Number 1999-01-2979]. Vancouver, B.C.

Patankar, M., & Taylor, J. (1999b). *Corporate aviation on the leading edge: systemic implementation of macro-human factors in aviation maintenance* [SAE Technical Paper No. 1999-01-1596]. Presented at the SAE General, Corporate & Regional Aviation Meeting & Exposition, Wichita, KS.

Patankar, M., & Taylor, J. (2000). Human Resources Integration master Plan: A Response to Revolving Door Management. SAE Technical Paper 2000-01-2128. In *Proceedings of the SAE Advances in Aviation Safety Conference & Exposition*. Daytona Beach, FL.

Patankar, M., & Taylor, J. (2001a). Analysis of organizational and individual factors leading to maintenance errors [SAE Technical Paper Number 2001-01-3005]. Warrendale, PA: Society of Automotive Engineers.

Patankar, M., & Taylor, J (2001b). Effects of MRM programs on the evolution of a safer culture in aviation maintenance. In R. Jensen (Ed.) *Proceedings of the Eleventh International Symposium on Aviation Psychology* at Ohio State University, Columbus, OH.

Patankar, M., & Taylor, J. (2002). Posterior probabilities of causal factors leading to unairworthy dispatch of a revenue flight. *Journal of Quality in Maintenance Engineering*. Bradford, U.K.: MCB Publishing.

Patankar, M.S., & Taylor, J.C. (2004). *Risk management and error reduction in aviation maintenance*. Aldershot, U.K.: Ashgate Publishing Limited.

Patankar, M., Taylor, J., & Goglia, J. (2002). Individual professionalism and mutual trust are key to minimizing the probability of maintenance errors. In *Proceedings of the Aviation Safety & Security Symposium*. Washington, D.C.

Phillips, D. (2002, December). *Improper use of tape to fix wings may lead to FAA fine for United.* Retrieved April 9, 2003 from http://www.washingtonpost.com/ac2/wpdyn?pagename=article&node=&contentId=A5484-2002Dec3¬Found=true.

Predmore, S.C., & Werner, T. (1997). Maintenance human factors and error control. In *Proceedings of the 11th FAA Meeting on Human Factors*

Issues in Aircraft Maintenance and Inspection. DOT/FAA/AM, Office of Aviation Medicine, Washington, DC, pp. 79-89.

Rankin, B., & Allen, J. (1996, April-June). Boeing introduces MEDA, Maintenance Error Decision Aid, *Airliner*, 20-27.

Robertson, M., Taylor, J., Stelly, J., & Wagner, R. (1995). A systematic training evaluation model applied to measure the effectiveness of an aviation maintenance team training program. In *Proceedings of the Eighth International Symposium on Aviation Psychology*. Columbus, Ohio, The Ohio State University, 631-636.

Sexton, J.B., & Klinect, J.R. (2001). The link between safety attitudes and observed performance in flight operations. In *Proceedings of the Eleventh International Symposium on Aviation Psychology*. Columbus, OH: The Ohio State University.

Taylor, F.W. (1911). *The principles of scientific management*. New York: Harper & Row.

Taylor, J. (1991). Maintenance organization. In Johnson, W.D., Drury, C.G., Taylor, J.C., & Berniger, D. (Eds.) *Human Factors in Aviation Maintenance. Phase 1: Progress Report* DOT/FAA/AM-91/16. Washington, DC: U.S. Dept of Transportation Federal Aviation Administration.

Taylor, J. (1995). Effects of communication & participation in aviation maintenance. In *Proceedings of the Eighth International Symposium on Aviation Psychology*. Columbus, Ohio: The Ohio State University, 472-477.

Taylor, J. (1998). Evaluating the effectiveness of Maintenance Resource Management (MRM). In *Proceedings of the Twelfth International Symposium on Human Factors in Aircraft Maintenance and Inspection*. Gatwick, UK, 85-99.

Taylor, J. (1999). Some effects of national culture in aviation maintenance. SAE Paper 1999-01-2980. In *Proceedings of the SAE Advances in Aviation Safety Conference*. Daytona Beach, FL.

Taylor, J. (2000a). Reliability and validity of the Maintenance Resources Management, Technical Operations Questionnaire (MRM/TOQ). *International Journal of Industrial Ergonomics, 26,* 217-230.

Taylor, J.C. (2000b) *Evaluating The Effects Of Maintenance Resource Management (MRM)*. In Air Safety Report of Research Conducted under NASA-Ames Cooperative Agreement No. NCC2-1025 (SCU Project # NAR003). Santa Clara University.

Taylor, J.C., & Christensen, T.D. (1998). *Aviation Maintenance Resource Management: Improving communication.* Warrendale, PA: SAE Press.

Taylor, J.C., & Felten, D.F. (1993). *Performance by design: Sociotechnical systems in North America.* Englewood Cliffs, NJ: Prentice Hall.

Taylor, J., & Patankar, M. (2001). Four generations of MRM: An analysis of the past, present, and future generations of MRM programs. *The Journal of Air Transportation World Wide*, *6* (2), 3-32.

Taylor, J., & Robertson, M. (1995). *The Effects Of Crew Resource Management (CRM) Training In Airline Maintenance: Results Following Three Years Experience.* NASA Contractor Report 196696, Washington, DC: National Aeronautics and Space Administration.

Taylor, J., Robertson, M., & Choi, S. (1997). Empirical results of Maintenance Resource Management training for Aviation Maintenance Technicians. In *Proceedings of the Ninth International Symposium on Aviation Psychology.* Columbus, Ohio, The Ohio State University, pp. 1020-1025.

Taylor, J., Robertson, M., Peck, R., & Stelly, J. (1993). Validating the impact of maintenance CRM training. In R. Jensen (Ed.), *Proceedings of the Seventh International Symposium on Aviation Psychology.* Columbus, Ohio: The Ohio State University, 538-542.

Taylor, J., & Thomas, R. (2003a). *Evaluating Behaviorally Oriented Aviation Maintenance Resource Management (MRM) Training And Programs: Methods, Results, And Conclusions.* 2002 Report of Research Conducted under NASA-Ames Cooperative Agreement No. NCC2-1156 (Santa Clara University Project # NAR006), to NASA-Ames Research Center, Moffett Field, CA June 30, 2003.

Taylor, J., & Thomas, R. (2003b). The structure of trust in aviation maintenance. *International Journal of Aviation Psychology*, *13*(4), pp. 321-343.

Thomas, R.L., & Taylor, J.C. (2003c). *Evaluating MRM Programs: The Use of Company- and Department-Level Percentile Ranks in Evaluating MRM Programs.* June. Accessed at http://mrm.engr.scu.edu/ newtool.html on 3/31/04.

Westrum, R., & Adamski, A. (1999). Organizational factors associated with safety and mission success in aviation environments. In D. Garland, J.

Wise, & V. Hopkin (Eds.), *Handbook of aviation human factors* (pp. 67-104). Mahwah, NJ: Lawrence Erlbaum Associates, Publishers.

Wrightsman, L.S. (1974). *Assumptions about human nature: A social-psychological analysis*. Monterey, CA: Brooks/Cole.

Index

T